Technology of Skilled Processes

Basic Engineering Competences

Measuring and Marking Out

Editorial Panel

T Beaumont, C Eng, MIMechE, MBIM
Vice-Principal
Huddersfield Technical College

N A Butterworth, MSc, CGIA, CEng,
FIMechE, FIProdE

G H Farnworth, MScTech, PhD, CEng,
MIMechE, MIProdE

V Green, TEng(CEI), MBIM
Head of Department of Engineering Crafts,
Huddersfield Technical College

C Sutcliffe, OBE, MSc, CEng, MIMechE
Vocational Curriculum Services
City and Guilds of London Institute

**Published as a
co-operative venture
between
Stam Press Ltd**

and

**City and Guilds of
London Institute**

Technology of Skilled Processes 367-1

Section	Process	Section	Process
1	Observing Safe Practices	8	Joining
2	Moving Loads	9	Fabrication
3	Measurement and Dimensional Control (1)	10	Soft Soldering, Hard Soldering and Brazing
4	Marking Out	11	Fusion Welding
5	Work and Toolholding	12	Power Transmission
6	Removing Material	13	Assembly and Dismantling (1)
7	Forming	14	Interpreting Drawings, Specifications and Data

Syllabus

Basic Engineering Competences 201

Basic Engineering Technology
201-1-01
01 Industrial Studies
02 Observing Safe Practices
03 Moving Loads
04 Measurement and Dimensional Control (1)
05 Marking Out
06 Work and Toolholding
07 Removing Material
08 Joining
09 Interpreting Drawings, Specifications and Data
010 Assembly and Dismantling (1)

Basic Fabrication and Welding Technology
201-1-07
01 Forming
02 Fabrication
03 Soft Soldering, Hard Soldering and Brazing
04 Fusion Welding

Basic Maintenance Technology
201-1-09
01 Forming
02 Soft Soldering, Hard Soldering and Brazing
03 Power Transmission
04 Measurement and Testing of Electro-Mechanical Systems (1)

Science Background to Technology
201-1-04
01 Basic Physical Quantities, Electricity and Magnetism
02 Forces
03 Pressure
04 The Principles of Tool Construction; Materials Technology

SUPPORTING BOOKS

Book titles	Covering	Covering
Basic Engineering	**Syllabus** 367-1	Syllabus 201-1-01
Observing Safe Practices and Moving Loads	Section 1 and 2	02-03
Measuring and Marking Out	Section 3 and 4	04-05
Work and Toolholding, Removing Material	Section 5 and 6	06-07
Joining	Section 8	08
Interpreting Drawings, Specifications and Data	Section 14	09
Assembly and Dismantling (1)	Section 13	10
Fabrication and Welding		Syllabus 201-1-07
Forming	Section 7	01
Fabrication	Section 9	02
Soft Soldering, Hard Soldering and Brazing	Section 10	03
Fusion Welding	Section 11	04
Maintenance		Syllabus 201-1-09
Forming	Section 7	01
Soft Soldering, Hard Soldering and Brazing	Section 10	02
Power Transmission	Section 12	03
Science		Syllabus 201-1-04
Basic Physical Quantities, Electricity and Magnetism		01
Forces		02
Pressure		03
Principles of Tool Constructions; Material Technology		04

201 – Basic Engineering Competences
201-1-01 Basic Engineering Technology

04 Basic Competence in Measurement and Dimensional Control (1)
05 Basic Competence in Marking Out

The contents of this book have been designed to cover the requirements of the City & Guilds Basic Process Compentence Syllabus (367-1), sections 3 and 4. The contents of the components 04 and 05 of the City & Guilds Basic Engineering Technology Syllabus 201-1-01 are identical and thus equally covered by this book.

As listed, the heading references in this book are conform with those in the syllabus sections 3 and 4 od scheme 367-1. In the 201 scheme syllabus items are numbered sequently and prefixed with the component number, e.g. item 1 in syllabus 04 is 4.1.
Below, in brackets following the page numbers we give the 201 syllabus sequence numbers.

Contents Measuring and Marking Out

Introduction

This book is intended for those who are, or will be, doing a practical job in industry.

It is specially written for those who need their technology as a background to their work and as a means of adapting to changes in working practices caused by technological advance. Where words such as ''he'' or ''craftsman'' appear in this series, they are to be interpreted as ''he/she'', ''craftsman/woman''.

This new series of textbooks presents the technology in terms of competence rather than working from a conventional theoretical base, i.e. the material will help readers understand:
- the use of
- the change to
- the development of
- other uses of,

industrial process technology and skills.

This book has been compiled after a survey of the industrial skilled processes which form the nucleus of occupational schemes and pre-vocational courses of the City and Guilds of London Institute and a comparison with provisions elsewhere in Europe.

Three basic facts emerged:
- the technology is common to many different schemes though the contexts of applications are very different;
- the technology is being taught in a variety of workshops in a variety of exercises related to the immediate needs of students and their industries; these industrially-related exercises formed excellent learning tasks and provided clear motivation for students because of their immediate relevance;
- the technology is so well integrated with the 'first-task need' that students did not recognize its relevance to many other tasks they would be called upon to perform.

This book seeks to build on the learning tasks and to provide a means of learning and generalizing the technology, so that the immediate job is better understood and better done, new tasks using the same process technology are more quickly mastered and updating or retraining is easier and more effective.

The editors are grateful to staff at Southampton Technical College and Neill Tools Limited for their advice on specific matters in this text. They would welcome further constructive suggestions which should be addressed to:
Stam Press Ltd
Raans Road
Amersham
Bucks. HP6 6JJ

Australia	AEP, Blackburn (Melbourne)
België	Plantyn, Deurne (Antwerpen)
Belgique	Plantyn, Bruxelles
BRD	Stam, Köln
France	Casteilla/Educalivre, Paris
Great Britain	Hulton, Amersham
	Stam Press, Amersham
	Stanley Thornes, Cheltenham
Nederland	Educaboek, Culemborg
	Educa Int., Culemborg
	De Ruiter, Gorinchem
Suisse	Delta & Spes, Denges (Lausanne)

First published in Great Britain 1986
as a co-operative venture between Stam Press Ltd and the City and Guilds of London Institute

© Stam Press Ltd, Amersham, 1986

ISBN 0 85973 015 8

Printed in Holland.

Project Structure and Use of Syllabus Bank and Supporting Books

1   The TECHNOLOGY associated with a given industrial process is a common requirement, but the APPLICATIONS vary by occupation and task, so a distinction has to be made between:
   (a)   THE AIM of the process: eg. to bend, metals, to drill, etc.
   (b)   THE LEARNING and ASSESSMENT: related to the application(s) specific to the industry to which the candidate belongs or aspires, or to the context of scheme chosen as a basis of study.

2   The approach suggested for the learning and assessment of any process technology is as follows:

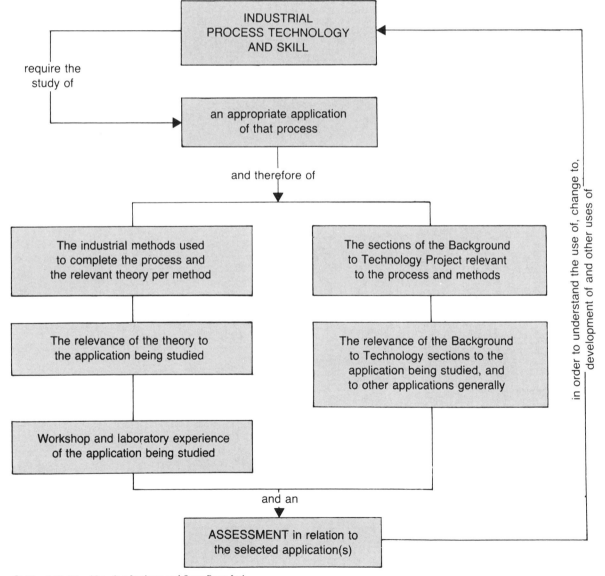

Part 1

1 Introduction to measurement

1.a Measurement as a comparator process

Measuring is a comparing process and, in measuring, a distinction can be made between comparing
- without using measuring tools or instruments
- using measuring tools or instruments.

Measuring equipment may be
- non-indicating
- indicating.

To show clearly the differences between the various methods of measurement (comparison) we will use a rectangular block as shown in Fig. **1.1.1** and describe it as a 'test piece'.

A number of workpieces are to be made to the same size as the test piece; for simplicity, we will restrict ourselves to measurement of the length only.

Comparing without measuring equipment
The test piece and the workpiece are placed on a horizontal plane in the same position (Fig. **1.1.2**). This method of measuring only permits us to state that
- the length of the workpiece is equal to the length of the test piece, **or**
- the length of the workpiece is not equal to the length of the test piece.

In practice, however, comparison of lengths without using measuring equipment is principally confined to the cutting off process (Fig. **1.1.3**).

1.a.i Comparing by using non-indicating equipment

Fixed checking tools
A checking tool may be made by using the test piece (Fig **1.1.4**). The internal size of the checking tool is equal to the length of the test piece. Each workpiece made may be checked using the checking tool. This tool is called a plate gauge.
It must be emphasised that here we can state only that
- the lengths are equal, **or**
- the lengths are not equal to the size of the gauge.

We can make the gauge as in Fig. **1.1.5** so that each

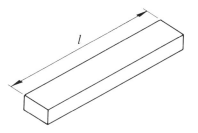

1.1.1   Test piece

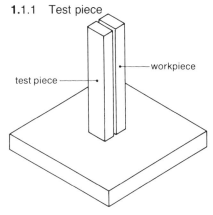

1.1.2   Comparing with a test piece

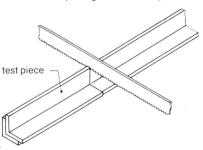

1.1.3   Cutting to length according to a test piece

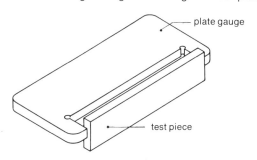

1.1.4   Comparing using a plate gauge

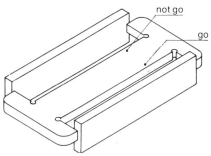

1.1.5   Comparing using a plate gauge with 'go' and 'not go' ends

workpiece made should enter the 'go' end, but not the 'not go' end. This permits a more accurate check than by the methods described earlier. Now we can state that

- the length is correct, **or**
- the length is not correct.

In practice this method of comparison is applied by using limit gauges (Fig. **1**.1.6a and b) which are quick and simple to use, when compared with direct measuring. They are extensively employed in large-scale production of components made to even high levels of accuracy, but it must be understood that they cannot be used to give the precise dimension of an item.

Adjustable measuring equipment
Adjustable measuring instruments are more versatile than those which are fixed, and can be used for checking a range of sizes (in Fig. **1**.1.7a an outside caliper is adjusted to the length of the test piece). The length of each workpiece made is compared with the dimension between the legs of of the caliper.

We can now state that

- the length is correct, **or**
- the length is not correct.

The following are practical examples of using adjustable measuring tools.

- Checking that the diameter of a shaft is uniform (Fig. **1**.1.7b).
- Checking that the angle between two surfaces is uniform along its length (Fig. **1**.1.8).

1.a.ii Comparing by using indicating measuring equipment

From a test piece
The length of the test piece is measured, e.g. with a steel rule. Next the length of each workpiece made is measured (Fig. **1**.1.9). In this manner we can determine

- the length of each block
- whether the lengths are similar
- whether the lengths are not similar
- how large any deviation is from the length required.

1.1.6a. Checking the diameter of shaft using a limit gauge: go side

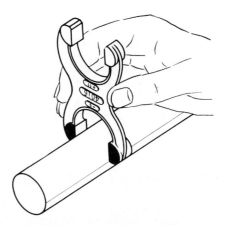

1.1.6b Checking the diameter of a shaft using a limit gauge: not go side

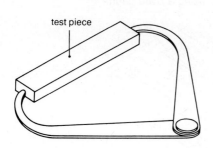

test piece

1.1.7a Comparing using outside caliper

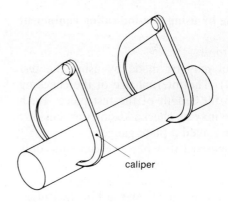

caliper

1.1.7b Checking parallelism of a shaft

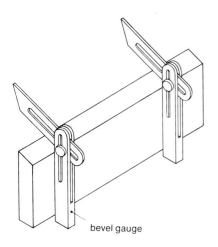

1.1.8 Checking an angle

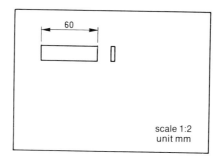

1.1.10 The drawing as a basis for measurement

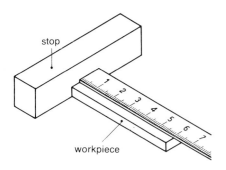

1.1.9 Measuring a length using an indicating measuring instrument

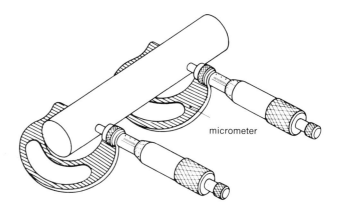

1.1.11 Measuring the diameter of a shaft

From a drawing

On a drawing, the length of a workpiece is specified by a dimension. We take the length of each workpiece made (Fig. **1.**1.9) and compare it with the length specified on the drawing (Fig. **1.**1.10). This enables us to state

- the lengths of the workpieces
- whether the lengths correspond to the dimension of the drawing
- whether the lengths do not correspond to the dimension on the drawing
- how large any deviation is from the dimension required.

Figs. **1.**1.11 and **1.**1.12 show some examples of indicating measuring equipment in use, but so far, discussion has been restricted to the length of the workpiece. Obviously, the width and the thickness of the workpiece can be measured in a similar way.

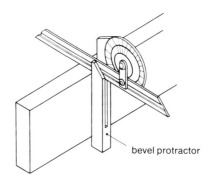

1.1.12 Measuring the angle between two surfaces

1.b Dimensional properties which can be measured

There is a wide variety of properties which can be measured. The most important ones are
- length
- flatness
- parallelism
- surface roughness
- angles
- profiles
- relative position of planes
- roundness and concentricity
- form accuracy.

1.b.i Length

Depending upon the shape of the workpiece the following features can be measured.
- The pitch — the distance between two points (Fig. **1**.1.13).
- The diameter of a shaft or a hole — the distance through the centre line, between two boundary lines (Fig. **1**.1.14). The distance is measured between a point on one line and a point on the other, or the distance is again measured between two points.
- The length, width and thickness of a block — each of these representing the distance between two surfaces (Fig. **1**.1.15). This distance is measured from a point in one plane to another point in another plane, i.e. the distance between two points. Any dimension in millimetres or parts of a millimetre relates to a distance. It is a

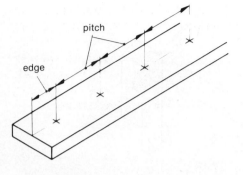

1.1.13 Pitch

length irrespective of the name given to it, e.g. width, thickness, diameter, radius, pitch, height, depth. Lengths are measured with a rule, a vernier caliper or a micrometer.

1.b.ii Flatness

Flatness may be checked in the following ways.
- By rubbing the worked surface in different directions over a surface plate on which a thin layer of marking medium has been applied, high points on the work are revealed by the marking medium as shining spots (Fig. **1**.1.16). We compare the flatness of the work with a flat plane (the surface plate).
- By holding a straight edge across the worked surface, deviations can be observed as a light gap (Figs. **1**.1.17 and **1**.1.18). In this manner we compare the flatness of the worked surface with a straight line (narrow plane). Checking should be carried out in different directions (Fig. **1**.1.19).

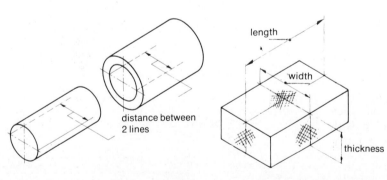

1.1.14   Outside and inside diameter   **1.1.15   Length, width, thickness**

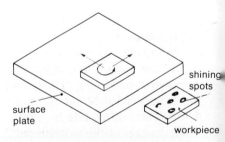

1.1.16   Inspection for flatness of a workpiece on a surface plate

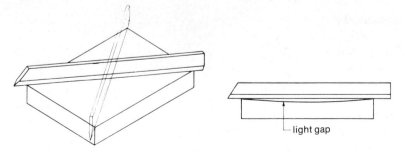

1.1.17 and 1.1.18   Checking the flatness of a workpiece using a straight edge

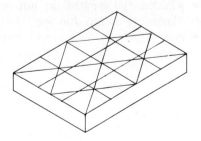

1.1.19   Checking in different directio

● By passing the stylus of a dial test indicator over the worked surface (Fig. **1.**1.20) deviations are indicated on the scale as lengths. In this manner we check whether several points on the worked surface are at equal heights from a flat plane. It is necessary to perform this check at different points over the whole surface area.

1.b.iii Parallelism

● Uniformity of the diameter of a shaft may be checked with a micrometer (Fig. **1.**1.21). What is measured is the distance between two points lying on an imaginary line through the centre of the shaft. Checking should be done at different places to make sure that the diameter is uniform along the length of the shaft.
● Uniformity of distance between two planes (surfaces) may be checked with:
a a dial test indicator (Fig. **1.**1.20)
b a micrometer

Measurements should be taken at different places on the workpiece to establish the extent of parallelism.

1.b.iv Surface roughness

The surface of a workpiece which has been machined or filed is always more or less rough. A surface filed with a coarse file is much rougher than when it has been produced on a grinding machine. Even a ground surface has a certain roughness. Fig. **1.**1.22 shows an enlarged cross-section of a worked surface. Such a roughness profile can be recorded with a surface roughness indicator (Figs. **1.**1.23* and **1.**1.24). In the determination of the roughness, a mean line is drawn through the recorded profile, in such a manner that the areas above the line are equal to the areas below it. This is done over the sampling length 1 (Fig. **1.**1.22). The average of the perpendiculars to the centre line denotes the average surface roughness over the sampling length. The most common international symbol for the measurement of surface roughness is the Ra number. It is the 'arithmetic mean' of the variation of the workpiece profile from the mean line (measured in μm or 0.000001 mm).

The roughness indicator shown in Fig. **1.**1.23 is now in use in workshops. This indicator registers the average roughness value in μm instantaneously as the stylus traverses the workpiece. The average roughness value can be read immediately from the frame on the indicator. The machine may be pre-set to the sampling length desired. The readings must be taken at several places on the workpiece surface, and the largest value is taken as the surface roughness of the workpiece.

* page 12

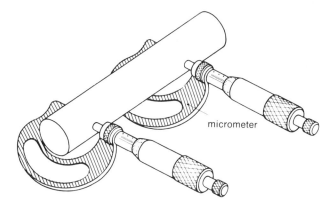

1.1.20   Checking the flatness of a workpiece using a dial test indicator

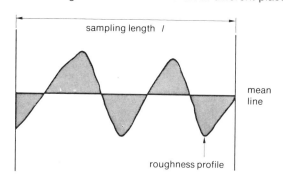

1.1.21   Measuring the diameter of a shaft at different places

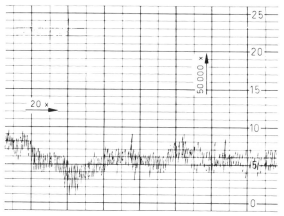

1.1.22   Representation of surface roughness

1.1.24   Graph of surface roughness

Roughness comparator scales may also be used in the workshop for checking surface roughness (Figs. **1.1.25** and **1.1.26**). The roughness of the comparator scale is compared with the roughness of the workpiece by scratching the surfaces with the fingernail (Fig. **1.1.27**). Visual comparison is also possible.

1.b.v Angles

Right angles (90°) are checked with a try-square (Fig. **1.1.28**). Any deviation is not registered in units with these tools but is shown as a gap between the square and the workpiece. However the universal bevel protractor shown in Fig. **1.1.29** may be used to measure right angles and other angles smaller and larger than 90°. Deviations can be read in degrees. A deviation of an angle can also be established in millimetres and fractions of a millimetre by using a cylindrical square and a dial test indicator on a stand (Fig. **1.1.30**).

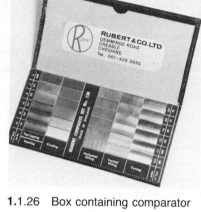

1.1.26   Box containing comparator scales

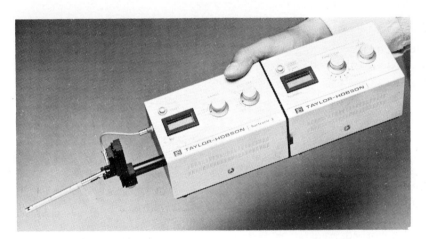

1.1.23   Surface roughness measuring instrument for workshops

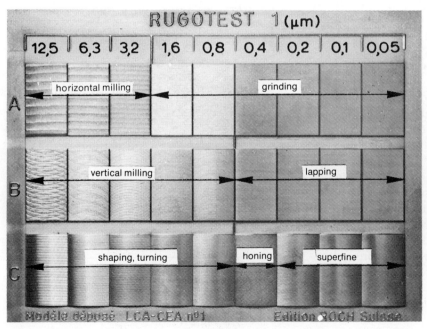

1.1.25   Roughness comparator scales for different operations

1.1.27   Checking surface roughness by comparison with roughness scales, using a fingernail

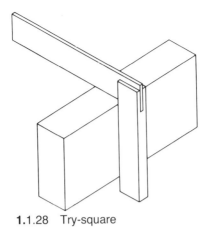

1.1.28 Try-square

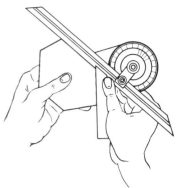

1.1.29 Universal bevel protractor
(reading in degrees)

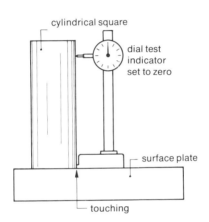

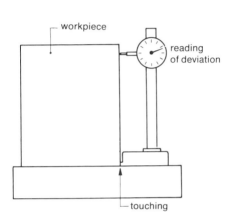

1.1.31 Radius gauges

1.1.30 Checking deviation of angle using a cylindrical square and a dial test indicator (deviation is expressed as a length)

1.b.vi Profiles

These are outlines which, for example, may appear as radii on workpieces and on screw threads. Circular profiles may be checked with radius gauges (Fig. **1.**1.31). The screw pitch gauge shown in Fig. **1.**1.32 is used to identify the types of thread. Optical projectors may be used for the accurate comparison of profiles (Fig. **1.**1.33). Projectors enlarge from 5 to 50 times. The profile of the object is compared with an accurately drawn profile, or with a template.

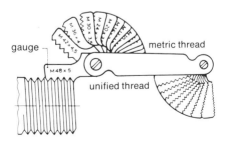

1.1.32 Screw pitch gauge

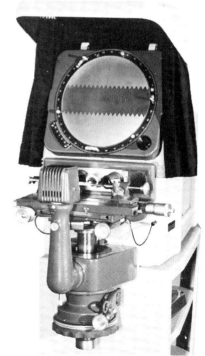

1.1.33 Optical comparator

1.1.34 Block level

1.1.35 Square block level (inside vial)

1.b.vii Relative position of planes

Plane surfaces are most frequently **either**
• horizontal and level, **or**
• vertical.

Horizontal planes
The horizontal position of a plane is checked with a block level (Fig. **1.**1.34) or a square block level (Fig. **1.**1.35). These are precision spirit levels. The deviation relative to a horizontal plane can only be measured with a spirit level which is provided with a scale giving the deviation in millimetres per metre of measured length (Fig. **1.**1.36). To check for a correct reading, the level is reversed through 180° and rechecked. The average of the two readings is the deviation relative to a horizontal plane.

Vertical Planes
A vertical plane can be checked by means of a square block level (Fig **1.**1.37).

Alignment testing
The following equipment can be used to check the alignment of a number of points.
• A wire and microscope.
• A lamp and sights (Fig. **1.**1.38).
• Special optical instruments (autocollimator or laser interferometer).

graduation 0.1 mm/m
bubble

1.1.36 Scale on the vial of a level (outside vial)

lamp
sight plate
eye

1.1.38 Lamp with sights

1.1.40 Uniform measurements of a hole which is not round

1.1.37 Checking a vertical plane using a square block level (outside vial)

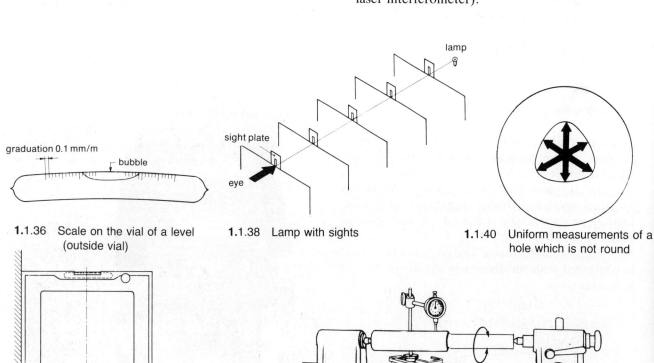

1.1.39 Measurement of roundness and run-out using a dial test indicator

1.b.viii Roundness and concentricity

The uniformity of diameter of a shaft can be checked with an outside caliper or micrometer. However, a shaft having a uniform diameter may
- not be straight
- rotate about two points which do not coincide with the centreline of the shaft.

Both cases are referred to as run-out. Run-out of a rotating shaft can be measured with a dial test indicator (Fig. **1**.1.39). The error is indicated by a dimension of length. Holes and shafts whose measurements show uniform dimensions may not be round. Fig. **1**.1.40 gives an example of such a hole.

1.b.ix Form accuracy

The accuracy of the combination of some or all of the dimensional properties discussed above is termed **form accuracy**. For example, a hexagonal prism has a close form accuracy when all six planes are perfectly flat, the planes run parallel in pairs and the planes are mutually square to the ends.
The number of measurements to be taken in checking whether a workpiece conforms to specifications of form accuracy or not, becomes larger when
- the workpiece is large
- the tolerances are small
- form accuracy has to be achieved under unfavourable machining conditions, e.g. instability of machines or poor tool and workholding conditions
- the stability of the workpiece is poor.

Influence of the manufacturing process on form accuracy
In the parallel filing of two planes, checking must be done more often than in parallel grinding on a surface-grinding machine. A surface-grinding machine usually provides such a high degree of reliability with regard to form accuracy that, providing it has been adjusted correctly, only one check is required.

1.c Agreed standards for measurement and dimensional control

1.c.i British and international agreements

In recent times much progress has been made in establishing standards, both nationally and internationally. For many years the British Standards Institution (BSI) has been the recognised authority for the preparation and publication of standards in the UK, whereas the International Organization for Standardization (ISO), as the name implies, sets standards world-wide. Metrication in Britain has provided the opportunity to align with ISO standards.

In this context it has been agreed that a uniform, accurately described unit of measurement should be used for the graduation of indicating measuring tools so that we can obtain the same dimensional number for the same measurement when readings are taken independently. Thus, confusion and misunderstanding are prevented. In addition, agreement has been reached on special requirements for the specification of workpieces, e.g. there are agreed standards for the form of the workpiece, the surface finish and the fits of different components. Whether or not a workpiece conforms to these requirements can be established by measurement.

This necessity for standardisation may be clarified by the following example. In years gone by, a cartwright (the maker of a product) who had to make a hay cart, knew for whom the cart was intended and for what purpose it was to be used. Parts which had to fit, were made to fit each other individually. Such a hay cart was built by a number of skilled men who were able to consult each other continuously (Fig. **1**.1.41). A worn part was replaced by a new one copied from the old part. Compare this handbuilt method of manufacture with the mass-produced modern motor car where 6000–8000 parts are involved, the parts coming from many different factories. Once assembled, the cars are then distributed to dealers in all parts of the world. When a worn part needs to be replaced it is exchanged for a new one, manufactured to the appropriate form accuracy, and is interchangeable with the old one, e.g. changing a wheel (Fig. **1**.1.42). Consultation between the manufacturers of the parts and the mechanics is no longer necessary. To ensure that every part fits and functions properly, it has to meet accurately specified

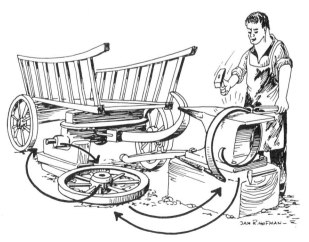

1.1.41

1.1.42

requirements, known as quality specifications. The working drawing is the usual medium for laying down quality specifications. In practice the conditions needed to satisfy these specifications are that

- the units of measurement and the quality specifications have to be accurately defined
- tools and machines have to be able to operate with the accuracy required
- the measuring tools used in the manufacture and inspection of the workpieces should also meet quality specifications
- measurement itself should conform to certain specifications
  (**Note** the influence of temperature on the results of measurement.)

1.c.ii Quantities and units

The characteristic of a quantity is that it can be expressed as a numerical value. An example of this would be the length, width and height of the building we are in. Here, the numerical values can be found by measuring the quantities with appropriate equipment. Conversely, we would find it difficult to express the attractiveness of the building quantitatively, or in numerical terms. Each quantity has its own unit of measurement; measuring is finding the number of units in a quantity.

The use of units in measurement
Length is a quantity whose unit is a metre. The length of the metre has been recorded on a platinum–iridium bar by two extremely thin lines (Fig. 1.1.43). This standard bar is not sufficiently accurate for certain measurements, and the metre is now redefined in terms of the wavelength of light. The metre is defined as 1 650 763.73 wavelengths, in a vacuum, of the orange radiation of the krypton-86 isotope.
When the metre length is transferred to a rule or tape, it enables us to measure, for example, a

football field. It is found that the rule has to be laid down 110 times to cover the length of the field. We then say that the football field is 110 times 1 metre or 110 × 1 metre long, therefore, the length of the football field is 110 metres.
The quantity of length is measured by comparing this quantity with the unit of length, thus measuring can be described as follows:

Measuring is comparing a quantity with a unit.

The quantity of time is measured by comparing it with the unit of time, which is the **second**. The unit of time is fixed as the time in which the hand travels from one division to the next, e.g. in a chronometer (Fig. 1.1.44).
The result of a measurement is a number followed by a matching unit. From the given examples the following statement can be made:

Quantity = number × unit

Frequently used quantities are
- length (thickness, height, distance, roughness)
- mass
- time
- velocity
- angle
- force
- pressure
- temperature
- electrical voltage
- electric current
- heat
- energy
- work

SI (Système International)
Each quantity should be given a certain unit, by which the quantity can be measured. To promote the international adoption of the same units the International System of Units was established in 1960 and is referred to as 'SI'. The characteristic of SI is that it has been designed in such a manner that a simple coherence exists between the various units.

Example The SI unit of length is the metre (m); the SI unit of time is the second (s). The unit of velocity is derived from length and time as the metre per second (m/s). The unit of mass is the kilogram (kg) (see also Fig. 1.1.45). The unit of force is the newton (N) and is derived from the units of mass, length and time.

1.c.iii Dimensional standards

Standardised systems of dimensional control developed slowly as interchangeability became an important factor in production (an interchangeable part can replace a similar part made from the same drawing). Work on the standardisation of screw threads in this country and the USA was followed by standards for limit gauges. These were replaced

1.1.43 International Prototype Metre (IPM): the standard metre until 1960

1.1.44 Chronometer

1.1.45 Standard kilogram

in the UK by a standard for limits and fits in engineering which was subsequently converted to the European metric standard and hence to a truly international system of limits and fits.

1.c.iv Technical terms and symbols

Terms and symbols are standardised in industry to provide a 'universal language' which is understood virtually anywhere in the industrial world. This is of particular value in the case of technical drawings and specifications, which need to have very precise definitions of the design and production requirements for a finished product. Details of technical terms and symbols can be obtained from BSI and ISO publications and specifications.

1.c.v Quality specifications for measuring instruments and equipment

Measuring instruments and equipment should have a higher accuracy than that required to meet the specified measurements for which they are used. Surface plates, micrometers, slip gauges, dial gauges, precision levels, etc., are covered by a range of BSI specifications which include recommended tolerances and often advice on the care and use of the equipment.

1.c.vi Environmental standards for measurement

Most materials used in engineering increase their linear dimensions when their temperature is increased, hence there is an internationally agreed standard temperature of 20°C for measuring purposes, and reference gauges (e.g. slip gauges, which may also be called gauge blocks are made to be of correct size at this temperature. It is not necessary for all measurements to be taken at this temperature, provided that the reference gauge and the work to be measured are made from similar materials, so that they expand and contract together. The main inspection department in an engineering company is often adjacent to the standards room, in which a range of high accuracy measuring equipment is housed. The temperature of the room is usually thermostatically controlled. The standards room is responsible for controlling the quality of production tooling equipment, including the measuring equipment and gauges used in the workshops and inspection department.

1.d The advantages of standards

By the application of suitable industrial standards for limits and fits to manufacturing operations, a degree of interchangeability can be achieved for many of the parts used in a firm's range of products. This can result in a large reduction in the quantity, and hence the cost, of stocks of raw materials and finished parts. It also reduces the cost of manufacture because increased output results from the larger batch sizes being produced in the workshops.

The acceptance of international standards for such components as fasteners, bearings and materials will again reduce the cost of manufacture, since standard components can be purchased at lower prices than special items. The larger quantities resulting from standardisation in the product range can achieve still further economies in the cost of 'bought out' components.

The quality of components produced in the workshops is influenced to a large extent by the condition of the tools, jigs and gauges used. All this equipment should be proved for accuracy before release to the workshops. It is essential that it should be regularly checked and where necessary re-calibrated to maintain the original accuracy.

2 Measurement of length

Types of equipment

The most commonly used equipment for measuring length is
- the rule
- the vernier caliper
- the micrometer
- the dial indicator.

This equipment is made to various patterns, depending on the purpose and the degree of accuracy required.

2.a Rules

2.a.i Sizes

Table 1.2.1 Rules

Measuring length (mm)	Width (mm)	Thickness (mm)
100–5000	13–60	0.3–12

2.a.ii Graduation and application

Most rules are provided with a graduation of 1 mm (Fig. **1.**2.1). They are used for measuring length with a tolerance of 1–3 mm.

2.a.iii Use of the rule

When handling the rule
- it should be positioned in the direction of the dimension to be measured (Fig. **1.**2.2)
- a datum block should be used whenever possible; it limits the possibility of reading errors to one place (Figs. **1.**2.3 and **1.**2.4)
- the eye should be positioned directly above the mark (Figs. **1.**2.5 and **1.**2.6).

2.a.iv Care of rules

Rules should be kept bright. Non-stainless steel rules need to be oiled regularly to prevent rusting. Regular inspection for accuracy is essential.

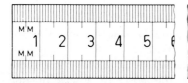

1.2.1   Rule

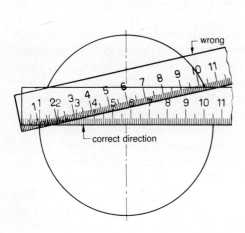

1.2.2   Correct and incorrect position of rule

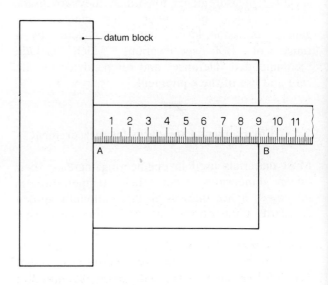

1.2.3   Reading error only at B

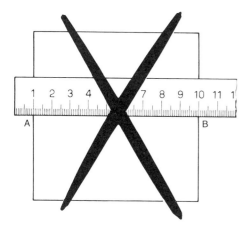

1.2.4   Reading errors likely at A and B

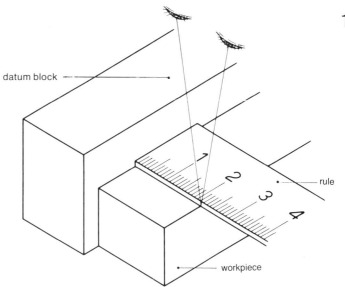

1.2.5   The eye should be positioned directly above the mark

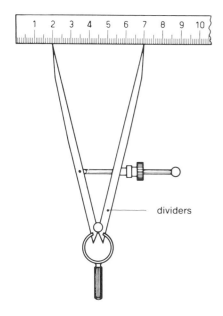

1.2.6   In transferring a size, the eye
should be positioned directly
above the mark

1.2.7   Measuring tape

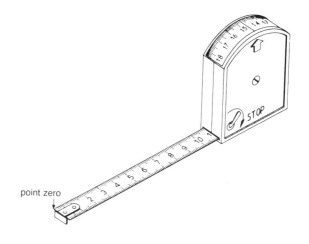

1.2.8   Measuring tape

2.a.v Similar tools

Measuring tapes of various kinds are used for measuring longer lengths (Figs. **1.**2.7 and **1.**2.8).

2.b Vernier instruments

2.b.i Vernier calipers

Fig. **1.**2.9 shows a vernier caliper with its different parts.

Dimensions

Table 1.2.2 Vernier calipers

Measuring range (mm)	Length between jaws (mm)
135–3000	80–500

Graduation reading

A vernier caliper can be read to:

0.05 mm with a vernier scale of ¹/₂₀ (Fig. **1.2.10**)

0.02 mm with a vernier scale of ¹/₅₀ (Fig. **1.2.11**)

Vernier calipers are usually made with both metric and imperial scales for use in the UK. Fig. **1.2.9** shows the upper beam and vernier scales reading in inches. With the 25 division scale shown, the caliper can be read to 0.001 in.

Application

Vernier calipers are used for length measurements with a tolerance of 0.2–0.5 mm, depending upon the measuring length and vernier scale.

Vernier scale

The use of a vernier scale is not limited to measuring tools. The same principle may be applied to graduated indexing rings on the slides of machine tools.

- *Vernier scale of ¹/₂₀*
  A length of 19 mm on the fixed scale corresponds with 20 divisions on the vernier scale, therefore, the distance between two division lines on the vernier scale is 19 mm ÷ 20 mm or 0.95 mm (Fig. **1.2.12**).
  The difference between 1 mm on the fixed scale and one division on the vernier scale is 1 mm − 0.95 mm or 0.05 mm.
  The difference for two vernier divisions is 2 × 0.05 mm or 0.1 mm. In this manner readings of 0.05 mm may be taken. Some examples of readings are shown in Figs. **1.2.13**–**1.2.16**. (Figs. 1.2.15 and 16 have been left for you to calculate and complete).

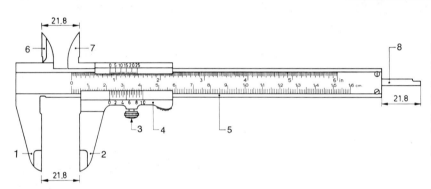

1 fixed jaw	for measuring
2 movable jaw	outside dimensions
3 clamping screw	
4 slide with vernier scale	
5 graduated beam	
6 fixed jaw	for measuring
7 movable jaw	inside dimensions
8 depth gauge	

1.2.9 Vernier caliper set to read 21.8 mm

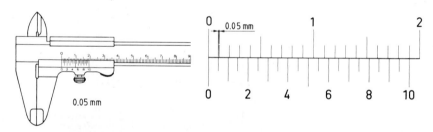

1.2.10 Vernier caliper with ¹/₂₀ vernier scale

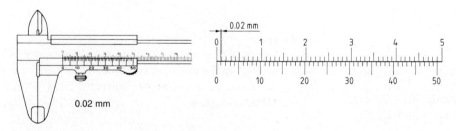

1.2.11 Vernier caliper with ¹/₅₀ vernier scale

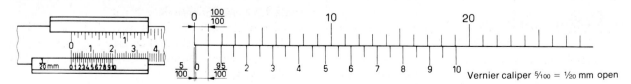

1.2.12 Vernier scale ¹/₂₀

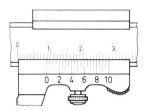 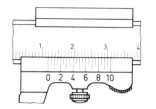 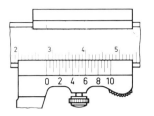 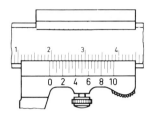

1.2.13 8 mm + 0.85 mm = 8.85 mm

1.2.14 11 mm + 0.95 mm = 11.95 mm

1.2.15 ... + ... = ...

1.2.16 ... + ... = ...

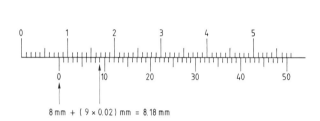

8 mm + (9 × 0.02) mm = 8.18 mm

1.2.17 Vernier scale ¹/₅₀

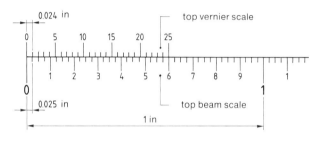

1.2.19a Imperial scale reading to 0.001 in.

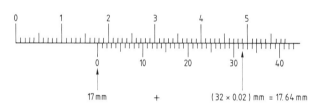

17 mm + (32 × 0.02) mm = 17.64 mm

1.2.18 Vernier scale ¹/₅₀

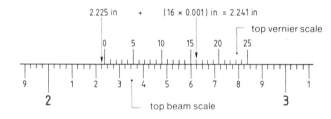

2.225 in + (16 × 0.001) in = 2.241 in

1.2.19b Imperial scale reading 2.241 in.

● *Vernier scale of ¹/₅₀*

A length of 49 mm on the fixed scale corresponds with 50 divisions on the vernier scale, therefore the distance between two division lines on the vernier scale is 49 mm ÷ 50 mm or 0.98 mm.

The difference between 1 mm on the fixed scale and one division on the vernier scale is 1 mm − 0.98 mm or 0.02 mm.

The difference for two vernier divisions is 2 × 0.02 mm or 0.04 mm.

The difference for three vernier divisions is 3 × 0.02 mm or 0.06 mm, thus, readings of 0.02 mm can be taken. Some examples of readings are given in Figs **1.2.17** and **1.2.18**.

● *Imperial vernier scale with 25 divisions*

A length of 0.6 in on the fixed scale corresponds with 25 divisions on the vernier scale, therefore the distance between two division lines on the vernier scale is 0.6 in ÷ 25 or 0.024 in (Fig. **1.2.19a**).

The difference between one division on the fixed scale and one division on the vernier scale is 0.025 in − 0.024 in or 0.001 in.

The difference between two divisions is 2 × 0.001 in or 0.002 in, thus, readings of 0.001 in can be taken. An example of a reading is given in Fig. **1.2.19b**.

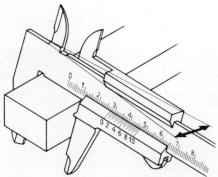

1.2.20a Make sure that measurement is with the flats of the jaws; check this by rocking the caliper gently

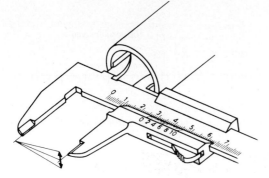

1.2.20b Insert the jaws as far as possible; check the proper position of the jaws by gentle rocking as shown

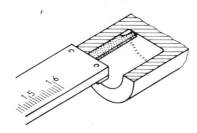

1.2.21 Recess of depth gauge should bridge the radius

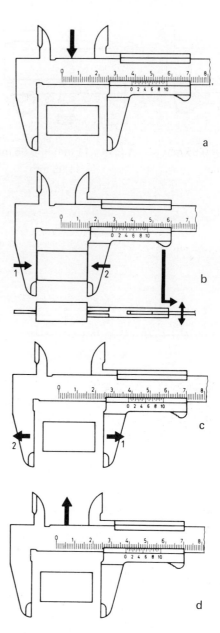

1.2.22 How to use a vernier caliper
- a open the jaws and bring the caliper over the workpiece
- b push the jaws against the work and take a reading
- c open the jaws
- d remove the vernier caliper

Use of the vernier caliper

In measuring a workpiece, the vernier caliper should be held parallel to the dimension to be measured (Fig. **1.2.20a** and b). The presence of radii and burrs should be taken into account (Fig. **1.2.21**). In addition, rubbing the jaw faces over the work should be avoided. The procedure is described in Fig. **1.2.22**. When reading the vernier, the eye should be positioned as when reading a rule. The pressure applied should be constant and light.

Care of vernier calipers

A vernier caliper has to be kept clean and bright. Regular application of oil after cleaning will prevent corrosion and ensure smooth functioning of the slide.

2.b.ii Types of vernier calipers

Many types of vernier calipers are available.
- The universal vernier caliper (Fig. **1.2.23**).
- The electronic digital caliper (Fig. **1.2.24**).
- The vernier caliper with a swing jaw (Fig. **1.2.25**).
- The vernier caliper with a hollowed jaw (Fig. **1.2.26**).
- The hole distance vernier caliper (Fig. **1.2.27**).
- Depth gauge (Fig. **1.2.28**).
- Vernier height gauge (Fig. **1.2.29**).

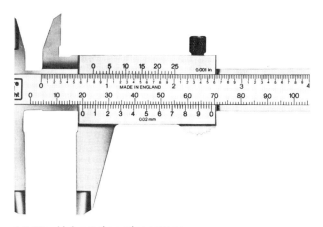

1.2.23   Universal vernier caliper

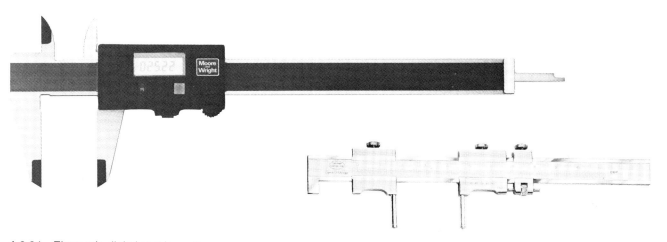

1.2.24   Electronic digital vernier caliper

1.2.27   Hole distance vernier caliper

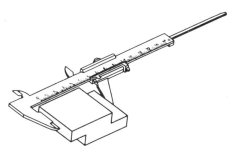

1.2.25   Vernier caliper with a swing jaw

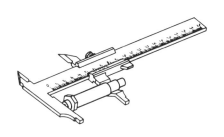

1.2.26   Vernier caliper with a hollowed
         jaw

1.2.28   Depth gauge

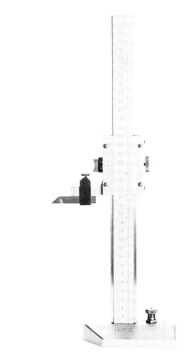

1.2.29   Vernier height gauge with
         scriber

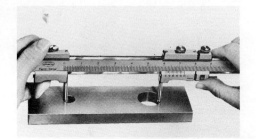

1.2.30 Hole distance vernier caliper

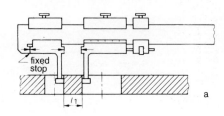

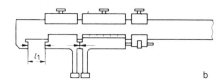

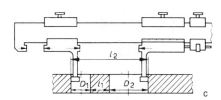

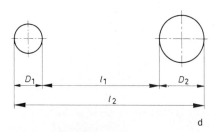

1.2.31 Use of hole distance vernier caliper
   a measure the smallest distance l_1
      and fasten the right hand slide
   b move the left hand slide against the
      right hand slide and fasten it
   c loosen the right hand slide and
      measure the largest distance l_2
   d total length $L = 2l_1 + D_1 + D_2$

2.b.iii Universal vernier caliper

Application
The universal vernier caliper may be used to measure
- outside dimensions
- inside dimensions
- depths.

For use in the UK, universal vernier calipers are mostly made with both metric and imperial scales (Fig. **1.2.23**). In this case, the upper beam scale and vernier read in inches: graduation usually enables reading to 0.001 in.

Care of the universal vernier caliper
The features which should be inspected regularly are
- the contact faces of the jaws for outside dimensions
- the contact faces of the jaws for inside dimensions
- the depth gauge.

2.b.iv Hole distance vernier caliper

Directions for use (Figs. **1.2.27** and **1.2.30**):
- Move the left hand slide to contact the fixed stop and measure the smallest distance l_1 between the holes (Fig. **1.2.31a**). Secure the right hand slide.
- Remove the caliper, push the left hand slide against the stop of the right hand slide (Fig. **1.2.31b**) and secure the left hand slide.
- Loosen the clamp screw of the right hand slide and measure the largest distance between the holes (l_2 in Fig. **1.2.31c**). Read the centre distance of the holes on the vernier scale of the right hand slide.

Example By following this procedure, the right hand slide has moved a distance, relative to the fixed stop, of

$$l_{\text{total}} = l_1 + l_2$$

(as shown in Figs. **1.2.31c** and **1.2.31d**)

$$l_2 = D_1 + l_1 + D_2$$

therefore

$$l_{\text{total}} = l_1 + l_2 + D_1 + D_2 = 2l_1 + D_1 + D_2$$

The beam is graduated twice full size (scale 2:1) so that the reading has to be halved to obtain a true measurement

$$\frac{2l_1 + D_1 + D_2}{2} = l_1 + r_1 + r_2$$

which is the centre distance desired.

1.2.32   Vernier height gauge with a
dial test indicator

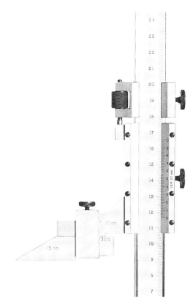

1.2.33   Vernier height gauge with digital readout

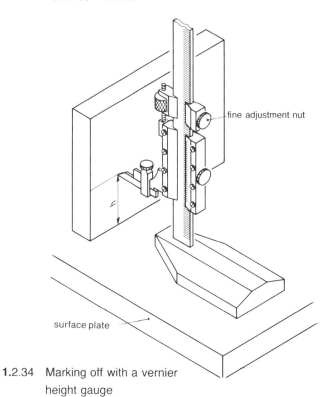

fine adjustment nut

surface plate

1.2.34   Marking off with a vernier
height gauge

1.2.34   Vernier height gauge with a scriber

2.b.v Vernier height gauge

Construction
The vernier height gauge shown in Fig. **1.**2.32 is
being used for measuring, and the one in Fig.
1.2.33 is also for measuring, but provides a digital
readout. The height gauge in Fig. **1.**2.34 is provided
with a scriber to be used for marking off.

Use of the vernier height gauge
The vernier height gauge may be used either for
measurement or for marking off on a surface plate.
When used for marking off, adjustment to the
height required may be carried out in two different
ways.

- The vernier height gauge is first adjusted to an
  approximate position and the clamp screw A is
  tightened. The correct size is then achieved by
  means of the fine adjusting nut B (Fig. **1.**2.35).
- The slip gauge is placed directly under the
  scriber (Fig. **1.**2.36). The sliding assembly is then
  locked to the beam by clamp screw C. The
  scribed line may then be produced.

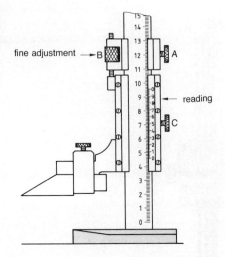

fine adjustment → B A

reading

C

1.2.35 Height adjustment; coarse adjustment, followed by fine adjustment (tighten screw A and turn nut B)

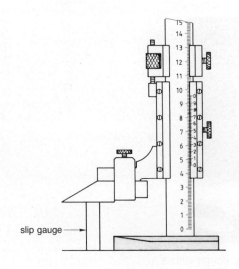

slip gauge

1.2.36 Height adjustment using a slip gauge under the scriber

2.c Micrometers

2.c.i Types of micrometer

Depending on their use we may distinguish between the following types
- the external micrometer (Fig. **1.**2.37) for outside dimensions
- the depth micrometer (Figs. **1.**2.38 and **1.**2.39) for measuring the depth of holes, recesses and slots
- the inside micrometer (Fig. **1.**2.40) for measuring inside dimensions.

Range of micrometers
Because of the errors which may accumulate in the manufacture of the spindle thread, the usual range of a micrometer and depth gauge is 25 mm.
To cater for varying dimensions which have to be measured, micrometers may be purchased with ranges of 0–25 mm, 25–50 mm, 50–75 mm and 75–100 mm. For dimensions larger than this micrometers designed for a range of 100–1000 mm are equipped with interchangeable anvils (Fig. **1.**2.41), permitting the same micrometer to be used for measuring dimension over this wide range.

Operation, graduation and reading
All micrometers operate on the same principle. In Fig. **1.**2.42 the work is positioned at A between an anvil and a spindle, the dimension is read at B.
In Fig. **1.**2.43 the spindle (2) has a screw thread of pitch 0.5 mm. Revolving the spindle one revolution will therefore change the distance between the anvil and the spindle face 0.5 mm. The thimble (7) in Fig. **1.**2.43 is graduated in 50 divisions. In this case the reading value, which is the distance the spindle moves when the thimble is turned one division is 0.5 mm ÷ 50 mm or 0.01 mm.

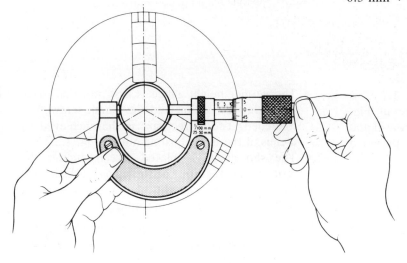

1.2.37 Measurement with an external micrometer

1.2.38 Depth micrometer

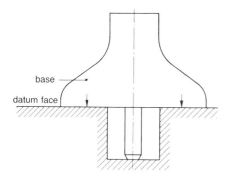

1.2.39 Measurement with a depth micrometer

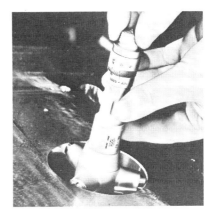

1.2.40 Measurement with an inside micrometer (three-point measurement)

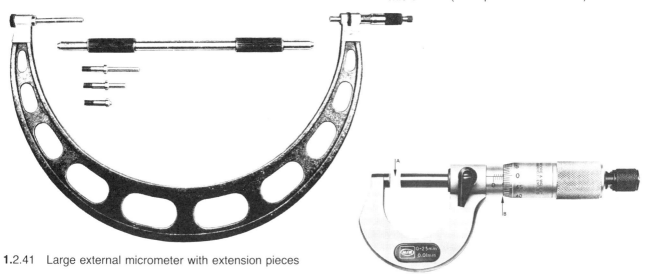

1.2.41 Large external micrometer with extension pieces

1.2.42 External micrometer

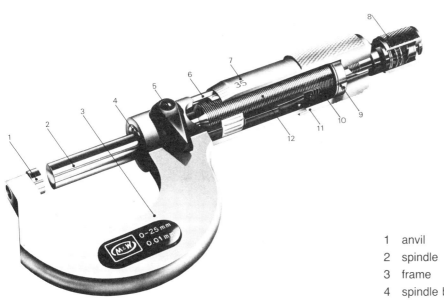

1.2.43 Construction of an external micrometer

1	anvil	7	thimble
2	spindle	8	ratchet
3	frame	9	thimble adjusting nut
4	spindle bearing	10	screw adjusting nut
5	locking lever	11	stem
6	barrel	12	spindle thread

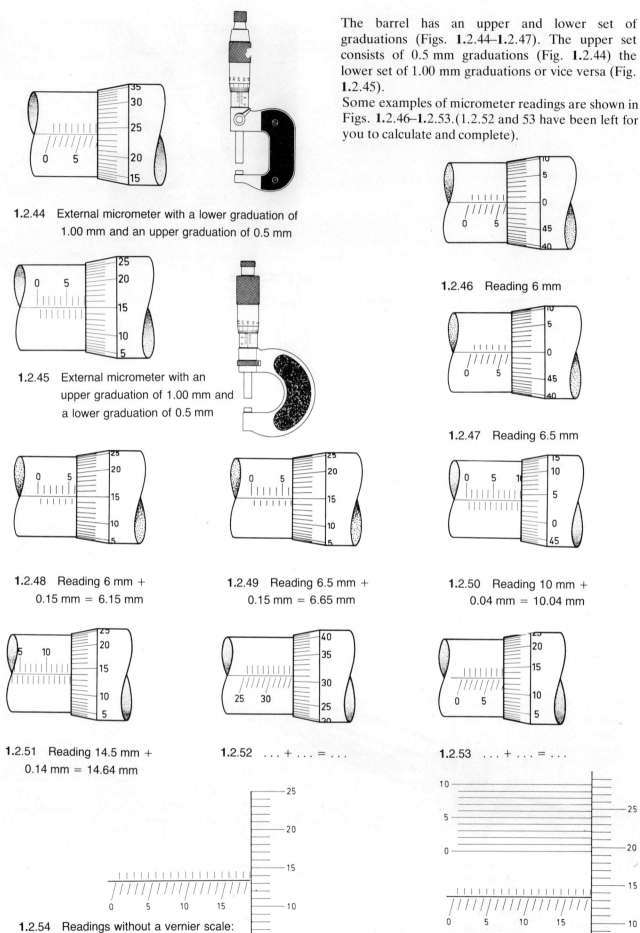

1.2.44 External micrometer with a lower graduation of 1.00 mm and an upper graduation of 0.5 mm

1.2.45 External micrometer with an upper graduation of 1.00 mm and a lower graduation of 0.5 mm

1.2.48 Reading 6 mm + 0.15 mm = 6.15 mm

1.2.49 Reading 6.5 mm + 0.15 mm = 6.65 mm

1.2.50 Reading 10 mm + 0.04 mm = 10.04 mm

1.2.51 Reading 14.5 mm + 0.14 mm = 14.64 mm

1.2.52 ... + ... = ...

1.2.53 ... + ... = ...

1.2.54 Readings without a vernier scale: size is between:
18.63 mm (18.5 + 0.13) and 18.64 mm (18.5 + 0.14)

The barrel has an upper and lower set of graduations (Figs. **1.**2.44–**1.**2.47). The upper set consists of 0.5 mm graduations (Fig. **1.**2.44) the lower set of 1.00 mm graduations or vice versa (Fig. **1.**2.45).

Some examples of micrometer readings are shown in Figs. **1.**2.46–**1.**2.53.(1.2.52 and 53 have been left for you to calculate and complete).

1.2.46 Reading 6 mm

1.2.47 Reading 6.5 mm

1.2.55 Reading with vernier scale:
18.5 mm + 0.13 mm + 0.004 mm = 18.634 mm

2.c.ii The external micrometer

Graduation and reading value

If a micrometer is provided with a barrel and a thimble graduation, as previously described, it can be read to 0.01 mm. Some micrometers have an additional vernier scale. A reading of 0.01 mm combined with a vernier scale of $\frac{1}{10}$ may give a reading to 0.01 mm ÷ 10 mm = 0.001 mm or 1 µm. Compare the readings in Figs. **1.**2.54 and **1.**2.55.

Application

External micrometers are used to measure outside dimensions with a tolerance of 0.01–0.1 mm.

The ratchet

The measuring force is brought about by turning the thimble. In order to obtain identical measurements, this force should be small and constant. For this purpose external micrometers are provided with a ratchet (Figs. **1.**2.43 and **1.**2.56). As soon as a certain force is reached, a friction spring or a ratchet mechanism comes into action, which prevents the spindle from being further tightened. This ensures that the **same force** can be applied by **any** user. The pressure, however, depends on the shape of the work (Fig. **1.**2.57). A difference in pressure may result in inaccuracy of the measurements.

Use of the micrometer

- Thoroughly clean the surfaces to be measured on the workpiece.
- Where possible a vernier caliper may be used to determine the approximate size in order to establish the range of the micrometer to be used (Fig. **1.**2.58).
- Clean the anvil and spindle faces of the micrometer.
- Set the micrometer approximately 0.5 mm larger than the vernier caliper reading.
- Move the spindle against the work, using the ratchet (Fig. **1.**2.56). Gently rock the micrometer to feel for proper contact of spindle and anvil.
- Take the reading. Look directly above the reading line. Neglecting to do this may result in a reading error of as much as 0.01 mm.
- Open the micrometer and keep it on a felt or rubber pad.
- When measuring, the entire faces of both anvil and spindle should be used. If there is no alternative to the use of the edges of the anvil and spindle, it should be remembered that the anvil and spindle faces may be convex and thus cause a measuring error.

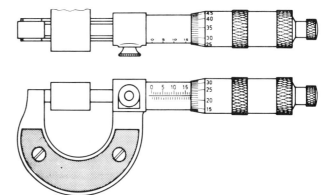

1.2.56   Use micrometer to determine the size accurately (reading 17.72 mm)

constant applied force

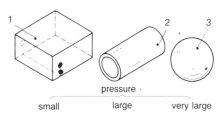

pressure

small   large   very large

1.2.57   A constant applied force gives a variable pressure, depending on the shape of the work and consequent areas of contact
1 full surface contact
2 line contact
3 point contact

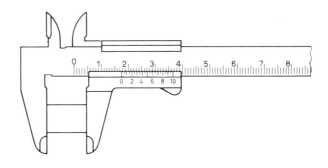

1.2.58   Use vernier caliper for approximation (reading between 17.7 and 17.8 mm)

1.2.59 Using a micrometer held in a stand

a

b

0 ... 25   0,001 0,01   0,1   mm

1.2.60 External micrometer with digital indication of 0.1 mm; the vernier scale gives readings accurate to 0.005 mm

1.2.61a External micrometer with digital indication accurate to 0.01 mm; an additional scale gives readings accurate to 0.0005 mm

1.2.61b Electronic digital micrometer

- Prolonged measuring may transmit the warmth of the hand to the frame in spite of the insulation pads. In such cases the use of a micrometer stand is recommended (Fig. **1**.2.59).
- Ensure that the temperature of the work is close to 20°C. In machining, for instance, it may otherwise be rejected due to shrinking after measurement.

Care of micrometers

Micrometers must be handled with care. Regular checking of the zero and some intermediate positions, with the aid of slip gauges, is necessary. Play in the screw and the adjustable nut can be taken up by tightening the nut. It is recommended that a micrometer should be set to the nominal size to be measured when the spindle is suspected of local wear. This should also be done in precision measuring, when the micrometer is set to either the nominal or the mean size with the aid of slip gauges.

2.c.iii Other types of micrometer

- Digital vernier micrometers (Fig. **1**.2.60). The hundredths of millimetres are displayed in the windows in multiples of 10. Fig. **1**.2.60 shows a reading of

$$12 \text{ mm} + 0.9 \text{ mm} + 0.03 \text{ mm} + 0.002 \text{ mm} = 12.932 \text{ mm}$$

The external micrometer shown in Fig. **1**.2.61a has the one-millimetre scale on the sleeve; the tenths and hundredths of millimetres are displayed in the windows. The scale graduation beside the figures on the right allows readings to be taken accurately to 0.005 mm. Reading to 0.001 mm has to be estimated. The micrometer in Fig. **1**.2.61a is set to read:

$$7 \text{ mm} + 0.25 \text{ mm} + 0.005 \text{ mm} = 7.225 \text{ mm}$$

- Electronic digital micrometers (Fig. **1**.2.61b). The digital display is automatically set to zero immediately the unit is switched on. At the touch of a button, instant zero set is obtained against a reference standard and therefore the instrument is self-calibrating. The direct comparison feature enables dimensional variations between like components to be accurately measured and indicated in the readout display as a plus or minus variation.
- Screw thread micrometer (Fig. **1**.2.62a and b) and standard micrometer with three measuring cylinders used to measure effective diameter of a screw thread (Fig. **1**.2.63).
- Large range micrometers (Fig. **1**.2.64).

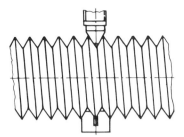

1.2.62a The spindle and anvil of a screw thread micrometer

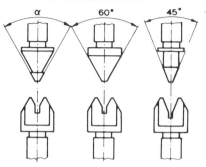

1.2.62b The shapes of the spindle and anvil are made to suit the thread form

left: inserts with short flanks for measuring (pitch) effective diameter

middle: inserts with full thread form for measuring (pitch) effective diameter

right: inserts with an angle of 15° or smaller for measuring minor diameter

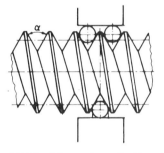

1.2.63 Measuring effective diameter of thread using three measuring cylinders

1.2.64 External micrometer for large diameters

2.c.iv Inside micrometers

There are two types:

- Inside micrometers for two-point measurements (Figs. **1**.2.65 and **1**.2.66).
- Inside micrometers for three-point measurements (Figs. **1**.2.67–**1**.2.70).

1.2.65 Inside micrometer for two-point measurement

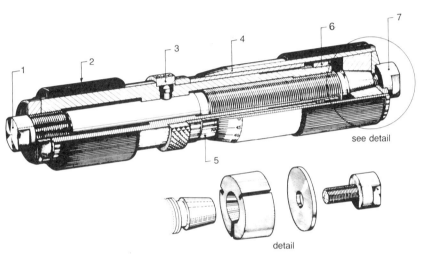

1.2.66 Principal parts of an inside micrometer

1	spindle	5	barrel
2	insulation	6	adjusting nut
3	locking device	7	spindle
4	thimble		

1.2.67 Three-point inside micrometer

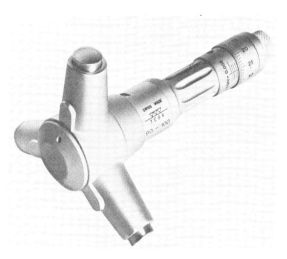

1.2.68 Three-point inside micrometer, for large dimensions

1.2.69   Three-point inside micrometer, range 10–100 mm

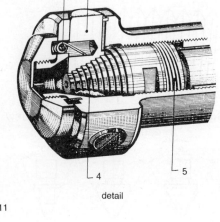

detail

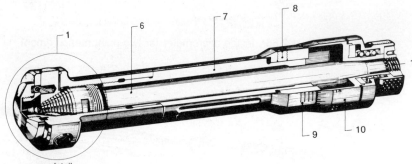

see detail

1.2.70   Principal parts of an inside micrometer

1   measuring head
2   measuring pin (3 off)
3   retraction spring
4   measuring screw thread
5   spindle travel thread
6   spindle
7   body
8   lock screw
9   barrel
10   thimble
11   ratchet

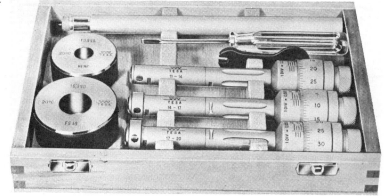

1.2.71   Inside micrometer set, with checking rings

Range of inside micrometers
Two-point inside micrometers have ranges of 25–30 mm and 30–35 mm. The range can be enlarged to 1500 mm by using extension pieces.
Three-point inside micrometers (Fig. **1**.2.68) have ranges of 6–8 mm, 8–10 mm, up to 200 mm.
Another type of three-point micrometer is shown in Fig. **1**.2.69. This has ranges of 10–12.5 mm, up to 100 mm capacity.

Graduation and reading
Two-point inside micrometers will give readings to 0.01 mm, and three-point inside micrometers will give readings to 0.005 mm.

Applications
● Two-point inside micrometers are often used for measurements ranging from 25 mm to 500 mm

and, depending on the measuring length, with a tolerance of 0.015–0.06 mm. The tolerance is increased for sizes larger than 500 mm.
● Three-point inside micrometers may be used for measurements ranging from 6 mm to 200 mm and a tolerance of 0.006–0.015 mm.

Care of inside micrometers
Inside micrometers are precision measuring instruments. They must be handled with care and checked regularly. Checking rings are provided with a set of instruments (Figs. **1**.2.71 and **1**.2.72). These checking rings enable any inaccuracy of the micrometer to be determined. Any deviation can be taken into account when measuring a workpiece. It should be added to or subtracted from the measurement taken, as appropriate.

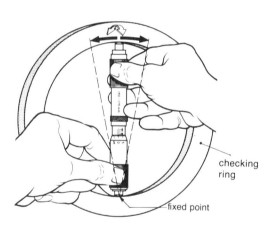

1.2.72a Checking three-point inside micrometer with a checking ring

1.2.72b Checking two-point inside micrometer with a checking ring

2.d Dial test indicators

Dial indicator dials vary in diameter from 25 mm up to approximately 60 mm.

Range and graduation

Table 1.2.3 Dial test indicators

Type of indicator	Range (mm)	Graduation value (mm)
dial test indicator (Fig. **1.**2.73)	up to 10.0	0.01
precision indicator	up to 1.0	0.001
precision indicator. (Fig. **1.**2.74)	approx 0.025	0.0005
lever type indicator (Figs. **1.**2.75 and **1.**2.76)	0.8	0.01
bore gauge with indicator (Fig. **1.**2.77)	depends on type of indicator	

1.2.73 Dial test indicator

1.2.74 Precision dial test indicator

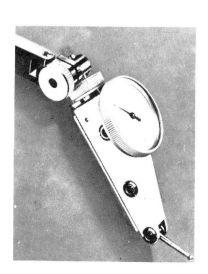

1.2.75 Lever type dial test indicator

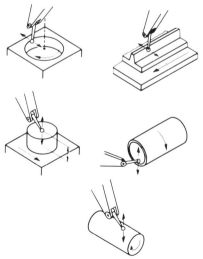

1.2.76 Application of a lever type dial test indicator

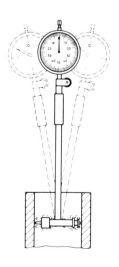

1.2.77 Bore gauge with dial test indicator

Application

The following general rules apply to dial test indicators.

- The smaller the range, the more precise will be the indicator.
- The smaller the distance travelled by the stylus, the more accurate the measurement.

The following values can be achieved (**Table 1**.2.4):

Table 1.2.4 Values on dial test indicators

Graduation value (mm)	Range up to (mm)	Tolerance depending on the measuring length (μm)
0.01	10.0	15–75
0.001	1.0	5–15
0.0005	0.025	2–5

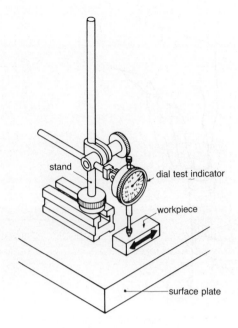

1.2.78   Measuring the parallelism of two planes

Measurement using dial test indicators

Dial test indicators are used to measure deviations

- in the parallelism of planes (Fig. **1**.2.78), shafts and holes
- in the roundness of shafts (Fig. **1**.2.79) and holes (Fig. **1**.2.80)
- in the relative position of planes to spindles (Figs. **1**.2.81–**1**.2.83).

In addition, checking for eccentricity may also be performed with a dial test indicator (Fig. **1**.2.84) Measurement with a dial test indicator may require that

- the work is moved while the dial test indicator is stationary (Fig. **1**.2.78)
- the dial test indicator is moved and the work is stationary (Fig. **1**.2.80)
- the work is revolved relative to the stationary dial test indicator (Fig. **1**.2.84)

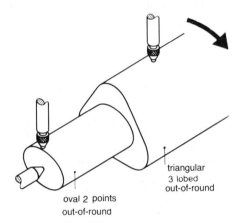

1.2.79   Measurement of ovality and 3 lobed out-of-roundness

In measuring the correct position of a spindle relative to a datum face

- the spindle with the dial test indicator (Fig. **1**.2.81) may be moved relative to the datum plane
- the spindle with the dial test indicator may be revolved relative to the datum plane (Fig. **1**.2.82)
- the dial test indicator may be moved along the datum plane relative to the spindle (Fig. **1**.2.83).

Dial test indicators may also be used to measure the size of a workpiece by comparison. For this purpose the indicator is set with the aid of a slip gauge (Fig. **1**.2.85). The reading on the dial test indicator (Fig. **1**.2.86) gives any difference in size between the slip gauge and the workpiece.

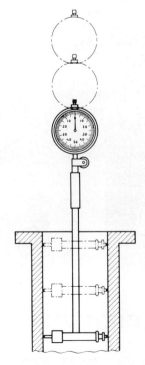

1.2.80   Measurement of deviations in diameter

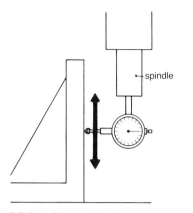

1.2.81 Measuring the vertical movement of a spindle relative to a plane

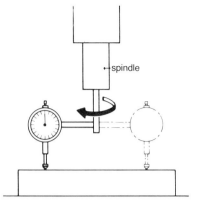

1.2.82 Measuring the squareness of a plane relative to a spindle

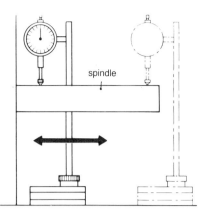

1.2.83 Measuring the parallelism of a spindle to a plane

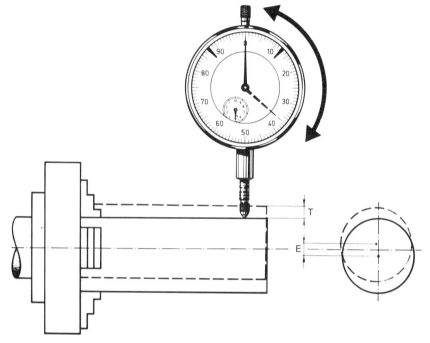

1.2.84 Measuring eccentricity

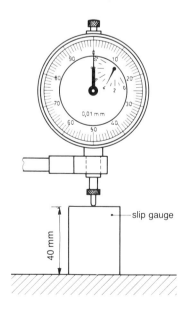

1.2.85 Setting a dial test indicator to a slip gauge

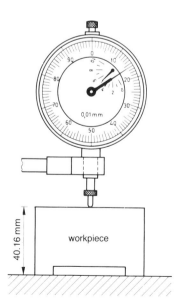

1.2.86 Accurate measurement of the dimension required, by comparison with the slip gauge setting

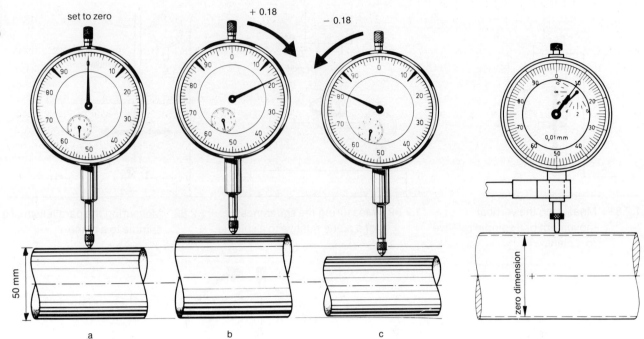

1.2.87 Measuring the diameter of shafts

1.2.88 Hand deflection to the right, indicates that the shaft is large than the zero dimension

Example 1 Indicator is set to an outside diameter of 50 mm (Fig. **1.**2.87a). The 0.01 mm dial indicator reads +0.18 mm (Fig. **1.**2.87b). The diameter is

$$50 \text{ mm} + 0.18 \text{ mm} = 50.18 \text{ mm}$$

Example 2 Indicator is set to an outside diameter of 50 mm (Fig **1.**2.87a). The 0.01 mm dial indicator reads –18 mm (Fig. **1.**2.87c). The diameter is

$$50 \text{ mm} - 0.18 \text{ mm} = 49.82 \text{ mm}$$

- In measuring outside dimensions a deflection to the right indicates an oversize and to the left an undersize (Figs. **1.**2.88 and **1**2.89).
- In measuring inside dimensions a deflection to the left indicates an oversize and to the right an undersize (Fig.s **1.**2.90 and **1.**2.91).

Directions for the use of a dial test indicator
- Clean the measuring surfaces and the stylus.

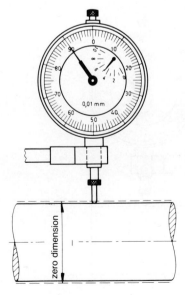

1.2.89 Hand deflection to the left, indicates that the shaft is smaller than the zero dimension

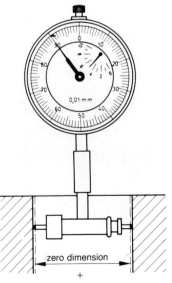

1.2.90 Hand deflection to the left, indicates a hole that is larger than the zero dimension

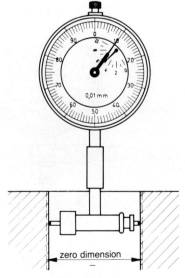

1.2.91 Hand deflection to the right, indicates a hole that is smaller than the zero dimension

- Allow for sufficient deflection of the hand to the right or left in setting the indicator (Figs. **1**.2.85 and **1**.2.86).
- Allow the stylus to descend on to the workpiece by the force of its spring.
- Lift the stylus when a recess or slot has to be crossed (Fig. **1**.2.92).
- The travel distance of the stylus should be as small as possible since the accuracy of measurement decreases as the gauging length travelled by the stylus increases.
- Clamping the indicator too tightly alters the measuring force applied (Figs. **1**.2.93 and **1**.2.94).

Care of dial test indicators
- Keep a non-watertight indicator away from coolants.
- Remove the indicator from the stand after use and put it away in a box or wooden case in a safe place.
- A magnetic base stand with a dial test indicator, when not in use, should be put away on a surface plate or tool locker, otherwise it may fall over owing to the effect of the weight of the indicator on the small base.
- Handle dial test indicators with care, even when they are shockproof.

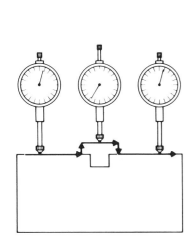

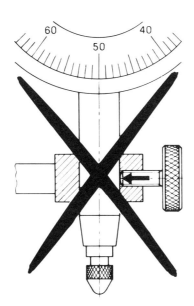

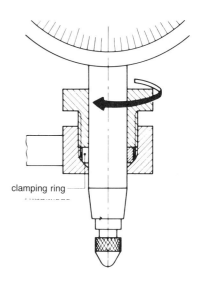

1.2.92   Lift the stylus across the slot

1.2.93   Bad clamping; indicator too tightly clamped on one side; the stylus may jam

1.2.94   Good clamping; indicator correctly clamped all round

3 Angle measuring equipment

Types of equipment

This includes equipment for measuring and checking position, planes and clearance.

3.a Squares

3.a.i Dimensions

The dimensions of squares refer to the lengths of the beam and the blade (Figs. **1.**3.1–**1.**3.4). A square with a stock length of 50 mm and a blade length of 75 mm is designated 75 × 50.

3.a.ii Applications

For use in engineering, squares are usually made to appropriate BSI standards, being graded as follows
- grade B — for normal workshop use, for checking and marking out right angles
- grade A — precision squares made for checking and inspection work
- grade AA — reference squares for checking other squares, used mainly in tool and standards rooms.

Squares give no indication of the amount of angular deviation in a work piece.

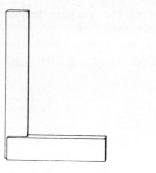

1.3.1 Stock try-square

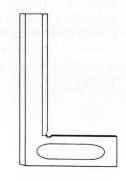

1.3.2 Bevel edge try-square

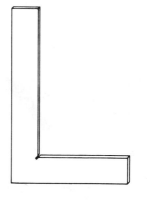

1.3.3 Try-square

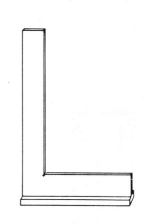

1.3.4 Try-square with datum edge

Table 1.3.1 Squares

	Stock try-square (Fig. **1.**3.1)		Bevel edge try-square (Fig. **1.**3.2)		Try-square (Fig. **1.**3.3) and try-square with datum edge (Fig. **1.**3.4)	
Designation	75 × 50	600 to 400	50 × 40	300 to 200	75 × 50	2000 to 1000
	mm	mm	mm	mm	mm	mm
Blade						
length	75	600	50	300	75	2000
width	16	44	13	40	12	60
thickness	2	4.5	4	9	4	12
Stock						
length	50	400	40	200	50	1000
width	14	42	—	—	—	—
thickness	10	23	—	—	—	—

3.a.iii Care of squares

Clean the square after use and oil it with a thin film of acid-free oil. Put precision instruments like bevel edge squares away in a box or tray (Fig. **1.**3.5).

3.a.iv Similar equipment

There are also instruments for checking other than right angles. Some of these are shown in Fig. **1.**3.6.

3.b Cylindrical and prismatic squares

3.b.i Sizes

Cylindrical squares (Fig. **1.**3.7) are manufactured in the sizes shown in table **1.**3.2 opposite.

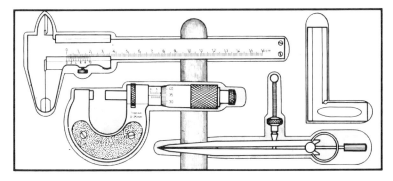

1.3.5   Storage of precision equipment

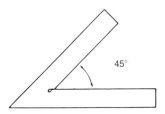
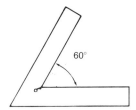
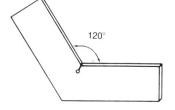

1.3.6   Tools for checking angles of 45°, 60° or 120°

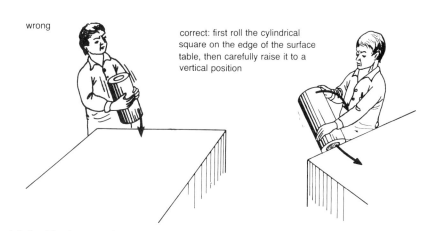

wrong

correct: first roll the cylindrical square on the edge of the surface table, then carefully raise it to a vertical position

1.3.7   Cylindrical square   **1.**3.8   Prismatic square   **1.**3.9   Placing a cylindrical square on a surface table

Table 1.3.2   Cylindrical squares

Height (mm)	Diameter (mm)	Mass (kg)
150	75	5.5
300	100	19.0
500	150	36.0
700	200	73.0

Fig. **1.**3.8 shows a similar type of square, but of prismatic form.

3.b.ii Application

A cylindrical square may be used to check the accuracy of try-squares and the squareness of the ends of a workpiece, or the parallelism between the faces of a workpiece. The component to be checked, and the cylindrical square, are placed on a surface table. In doing this, care must be taken to avoid damage to the edge of the bottom face of the cylindrical square (Fig. **1**.3.9).

3.b.iii Care of cylindrical squares

Precautions should be taken to ensure that the cylinder area and the base remain bright and undamaged. Cylindrical squares must be cleaned and oiled or greased after use.

3.c Angular measuring instruments

3.c.i Angles and the unit of angle

An angle is part of a circle. For the determination of the unit of angle, the circle is divided into 360 equal parts (Fig. **1**.3.10). $\frac{1}{360}$ part is called a degree, which is the unit of angle.

One degree can be divided into 60 equal parts called minutes.

$$1 \text{ degree} = 60 \text{ minutes or } 60'$$

By again dividing one minute into 60 equal parts, the second is obtained.

$$1 \text{ minute} = 60 \text{ seconds or } 60''$$

Thus

$$1 \text{ degree} = 60 \times 60 \text{ seconds} = 3600 \text{ seconds or } 3600''$$

This (angular) second has nothing to do with the second as a unit of time.

In Fig. **1**.3.11 the circle is divided into four equal right-angled parts by co-ordinates. A right angle measures 90°. In Fig. **1**.3.12 a right angle is divided into one angle of 20° and one of 70°. They are known as complementary angles, which together form a 90° angle.

A straight line is 180° = 2 × 90°. In Fig. **1**.3.13 a straight line (or straight angle) is divided into one

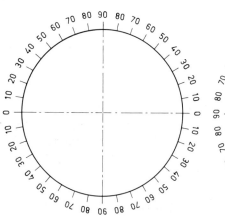

1.3.10   A circle is divided into 360°

1.3.11   A right angle and a straight angle

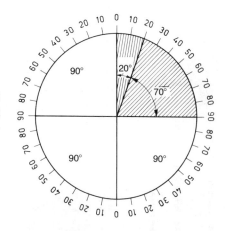

1.3.12   Complementary angles

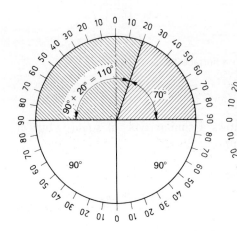

1.3.13   Supplementary angles

1.3.14   Supplement of an angle

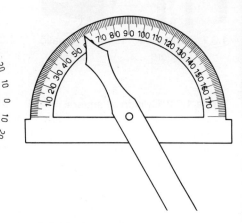

1.3.15   Protractor with indicator

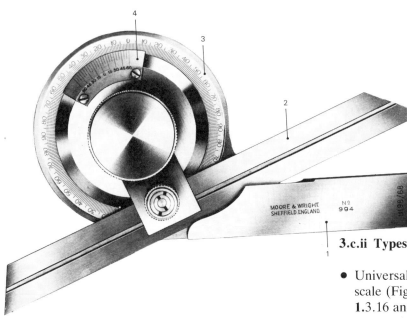

1.3.16   Universal bevel protractor
   1   blade
   2   stock
   3   graduated dial
   4   vernier scale

angle of 110° and one of 70°. These are
supplementary angles, that is, they supplement one
another to give 180°. The straight angle shown in
Fig. **1**.3.14 is divided into an angle of . . . . . and
one of . . . . . . They are one another's . . . . ?
(Fill in the missing detail)

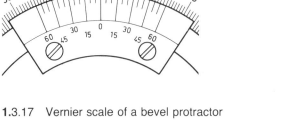

1.3.17   Vernier scale of a bevel protractor

3.c.ii Types of angular measuring instruments

- Universal bevel protractor without a vernier
  scale (Fig. **1**.3.15) and with a vernier scale (Figs.
  1.3.16 and **1**.3.17).Fig. **1**.3.16 shows the principal
  parts of a universal bevel protractor with a
  vernier scale. The vernier scale indicates to the
  left and right of zero (Fig. **1**.3.17).
- Optical bevel protractor (Fig. **1**.3.18). This type
  of protractor (Fig. **1**.3.18) has a very accurate
  dial graduation of 10′, applied by photographic
  processing. Readings are taken through the
  optical window which has a thirty-fold
  magnification.

From here, discussion will be confined to the
universal bevel protractor.

Protractor readings
The reading accuracy of a bevel protractor without
a vernier is 0.5° (Fig. **1**.3.15). Depending on the
graduation, the reading accuracy of a bevel
protractor with a vernier is
- 5′ for a 12 division vernier scale
- 2′30″ for a 24 division vernier scale.

12 division vernier scale
An angle of 23° is divided into 12 divisions on the
vernier scale, therefore, each division equals

$$^{23}/_{12}° = 1^{11}/_{12}° \text{ or } 1°55'$$

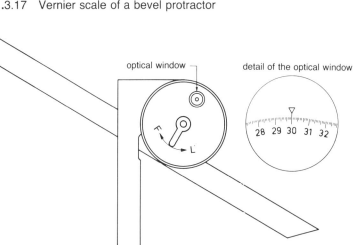

1.3.18   Optical bevel protractor

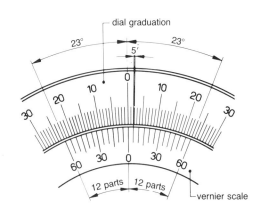

1.3.19   12 division vernier scale (reading
in 5′ increments)

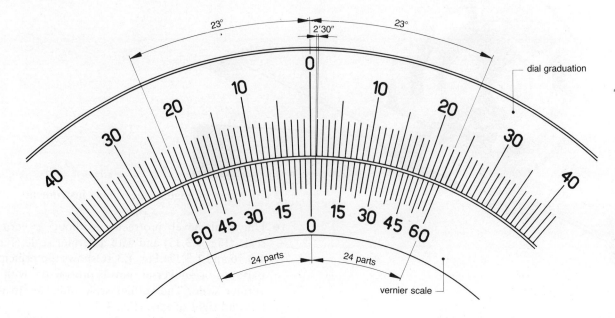

1.3.20   24 division vernier scale (reading in 2′30″increments)

The difference between 2° on the dial and one vernier division is

$$2° - 1°55' = 5' \text{ (Fig. 1.3.19)}$$

24 division vernier scale
Each division of the vernier scale here equals

$$^{23}/_{24}° = 57'30''$$

The difference between 1° on the dial and one vernier division is

$$1° - 57'30'' = 2'30'' \text{ (Fig. 1.3.20)}$$

Application
The manufactured maximum error for a good quality instrument will not exceed ±5′. This restricts the use of vernier bevel protractors to angular measurements with a tolerance of 0.5° = 30′ or larger, to ensure that the tolerance is not exceeded by the inaccuracy of the measuring instrument (Fig. 1.3.21).

Care of bevel protractors
Bevel protractors should be handled carefully. The

base on which the blade slides has to be kept clean and regular checking is necessary.

Reading a vernier bevel protractor
Figures 1.3.22–1.3.25 show a vernier bevel protractor in four different positions, each with the same reading. The figures illustrate that the reading refers to the angle between the stock base and one side of the blade. Since both the stock and the blade have parallel slides, the indicated angle (the reading) is always equal to the angle between the base of the stock and one side of the blade. The angle between the planes A_1 and L_1 (Fig. 1.3.23) is similar to that between the planes A_2 and L_2, i.e. the reading.
The vernier bevel protractors shown in Figs. 1.3.26–1.3.29 all have an identical reading. The angle we want to know may be equal to
• the reading (Fig. 1.3.26)
• 180° − the reading (Fig. 1.3.27)
• 90° + the reading (Fig. 1.3.28)
• 90° − the reading (Fig. 1.3.29)

The examples given in Figs. 1.3.22–1.3.29 mainly relate to a measurement where the stock and the work are resting on the same reference base. It will be clear that measurements may also be taken with the work held between the stock and the blade. One of the surfaces of the work then functions as a reference plane.
Figs. 1.3.30–1.3.33 give examples, with an explanation of where (right or left) the vernier value should be read for fractions of degrees, and how to do the necessary counting.
The measurement of angle α in Fig. 1.3.30 is an exterior angle and is said to be an outside measurement. Angle β in Fig. 1.3.30 is an interior angle, the measurement of which is an inside measurement.

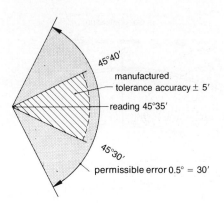

1.3.21   Relationship of workpiece tolerance and bevel protractor accuracy

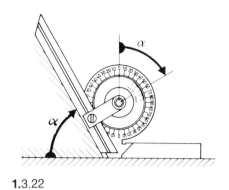

1.3.22

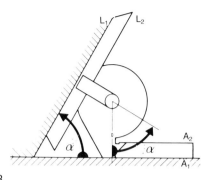

1.3.23

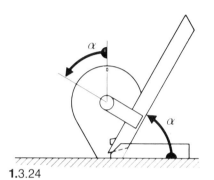

1.3.24

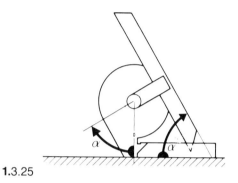

1.3.25

The same angle with different positions of the blade

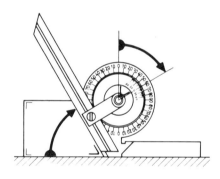

1.3.26   Angle required = reading; fractions of degrees to be read from vernier on the right

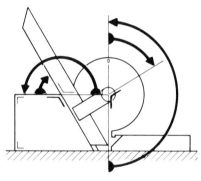

1.3.27   Angle required = 180° − reading; fractions of degrees to be read from vernier on the right

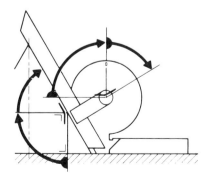

1.3.28   Angle required = 90° + reading; fractions of degrees to be read from vernier on the right

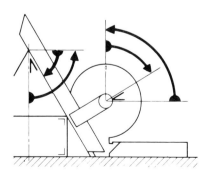

1.3.29   Angle required = 90° − reading; fractions of degrees to be read from vernier on the right

44

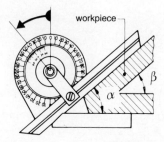

1.3.30 Acute angle: angle required =
the reading; use the left hand vernier

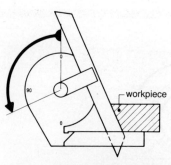

1.3.31 Obtuse angle: angle required is to be found by
additional counting (80 ≙ 100°, 70° ≙ 110°,
60° ≙ 120°) using the left hand vernier
(≙ means corresponds to)

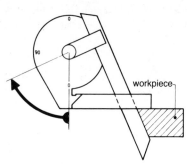

1.3.32 Acute angle: angle required =
the reading; use the right hand vernier

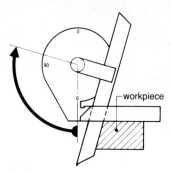

1.3.33 Obtuse angle: angle required is to be
found by additional counting; use the right
hand vernier

4 Accuracy of measurement

4.a Dimensional deviations

Deviations are the differences between true values, or sizes specified for features of a component and what is actually achieved in manufacture. These differences must be restricted, if the component is to be useful.

4.a.i-iv Reasons why they are allowed

In examining these we will restrict our discussion to length.
- The precise and absolute value of length cannot be measured, even with the most accurate measuring instrument.
- Technically it is impossible to attain the same level of accuracy with different production processes, e.g. casting, grinding, turning.
- A high degree of accuracy is not always necessary and using a higher degree than is needed will increase production costs.

There are therefore several reasons for allowing dimensional deviations. The size of the part and the allowable deviations must be specified on the drawing.

Nominal size
This is a dimension by which the size of a feature of a component is designated for convenience.

Limits
Limits (of size) are the high and low values of the maximum and minimum sizes permitted for the feature.

Tolerance
Tolerance is the difference between the maximum limit and the minimum limit of size.

Unilateral and bilateral tolerances
The dimension 40 ± 0.5 mm (Fig. **1**.4.1) permits deviation both above and below the nominal size and the tolerance is said to be *bilateral*. Had the tolerance been restricted to the nominal size $+ 0.5$ mm, or to the nominal size -0.5 mm, they would be *unilateral* tolerances.

Deviation
We can say that deviation is the difference between a limit and the nominal size.

Symmetrical and asymmetrical deviations
The size 40 mm ± 0.5 mm (Fig. **1**.4.1 and *Example 1*) has a positive deviation which is the same as the negative deviation, and therefore is symmetrical. An asymmetrical deviation is shown in the size $40^{+0.7}_{-0.3}$ where the low limit is 39.7 mm and the high limit 40.7 mm (*see Example 3*).

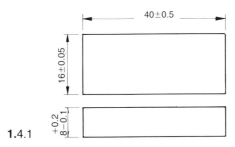

1.4.1

Example 1

Indicated length (Fig. **1**.4.1)	40 ± 0.5
Nominal size	40.0 mm
Low limit	39.5 mm
High limit	40.5 mm
Deviation	± 0.5 mm
Tolerance $= 40.5 - 39.5 =$	1.0 mm

Table 1.4.1 Metric units of length

Name	Multiple or decimal (metre)	Abbreviation (metre)	Symbol
1 kilometre	1000.0	10^3	km
1 metre	1.0	10^0	m
1 centimetre	0.01	10^{-2}	cm
1 millimetre	0.001	10^{-3}	mm
1 micrometre	0.000 001	10^{-6}	μm
1 nanometre	0.000 000 001	10^{-9}	nm

Table 1.4.2 Special angles and angle units

Name	Magnitude expressed in the unit			Symbol
	degree	minute	second	
circle	360	360×60	360×3600	$\varnothing$
right angle	90	90×60	90×3600	L
1 degree	1	1×60	1×3600	$1°$
1 minute	1/60	1	60	$1'$
1 second	1/3600	1/60	1	$1''$

Example 2

Indicated breadth	16 mm ± 0,05 mm

Nominal size	16.00 mm
Low limit	15.95 mm
High limit	16.05 mm
Deviation	± 0.05 mm
Tolerance = 16.05 − 15.95 =	0.1 mm

Example 3

Indicated thickness	8 mm	$\begin{array}{c} +\,0.2\,\text{mm} \\ -\,0.1\,\text{mm} \end{array}$

Nominal size	8.0 mm
Low limit	7.9 mm
High limit	8.2 mm
Positive deviation	+ 0.2 mm
Negative deviation	− 0.1 mm
Tolerance = 8.2 − 7.9 =	0.3 mm

Mean dimension
The mean dimension lies halfway between the high and the low limit.

Tolerances and the manufacturing method
The faces A and B of the block shown in Fig. **1.4.2** have to be filed. The shaft shown in Fig. **1.4.3** has to be machined in a lathe. Both workpieces have a nominal size of 50 mm. The limits in both cases are 50.1 mm and 49.9 mm. As soon as the size of the block in Fig. **1.4.2** reaches 50.1 mm, it conforms to the specified tolerance. Further finishing until the mean size is attained is not necessary and it only makes the workpiece more expensive. In the case of the machined shaft (Fig. **1.4.3**) it should take the same time to finish the work to a final size of 50.1 mm as to a size of 50.0 mm.
In general, when considering the tolerance and the mean size it can be said that

- the dimension is correct when it lies between the specified limits
- it is recommended that the mean size should be approached as closely as possible, provided that this does not involve additional time.

4.b The significance of accuracy in measuring

Accuracy is meaningful when, in a given situation, we have regard to

- the magnitude of the tolerance; in other words the boundaries of the permissible deviation of the measured dimensions from the nominal size
- the approximation to the mean size under practical manufacturing conditions; this may require the avoidance of accumulative errors, for example by using selective assembly methods.
- the relationship between the tolerance allowed and the nominal size. A tolerance of +0.1 mm on a nominal size of 2400 mm is more accurate than a tolerance of +0.01 mm on a nominal size of 6 mm.
- The accuracy of measuring instruments. More accurate measuring instruments have to be used as the permissible deviations to be measured become smaller. Table **1.4.3** gives the measuring accuracy of four length measuring instruments.

4.c Factors affecting the accuracy of measurement

The principal factors which affect the accuracy of measurement are

- the temperature
- inaccuracy of measuring equipment
- reading errors
- application of force when measuring
- the skill of the craftsman
- cleanliness of workpiece and measuring equipment
- deflection of measuring equipment and workpiece.

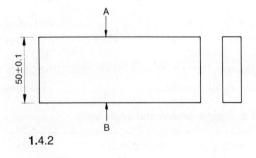

1.4.2

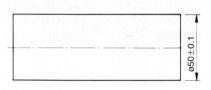

1.4.3

Table 1.4.3 Accuracy of some measuring instruments

Tolerance on workpiece is equal to or larger than	Instruments to be used
mm	
1.0 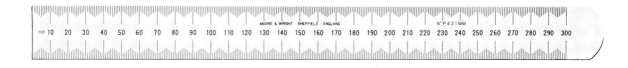	rule
0.25 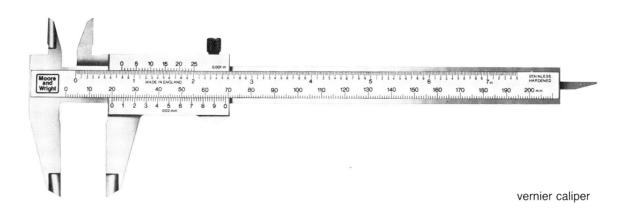	vernier caliper
0.03 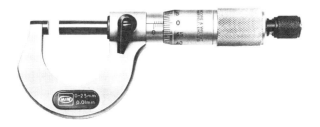	micrometer
0.02 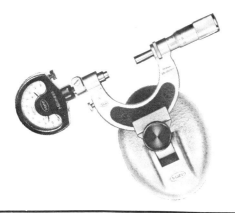	dial indicator

48

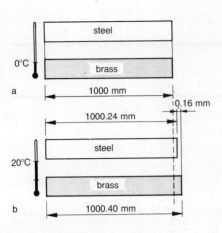

1.4.4 Effect of temperature on the expansion in length

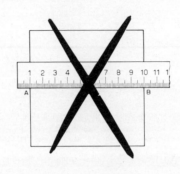

1.4.5 Reading errors at A and B

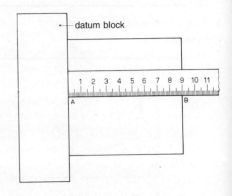

1.4.6 Reading error at B only

4.c.i Temperature

Temperature and material of the measuring equipment

A steel bar and a brass bar have an equal length of 1000 mm between the end faces at a temperature of 0°C (Fig. **1.4.4a**). When brought into a room at 20°C, these bars will reach the same temperature and consequently expand (Fig. **1.4.4b**). The length of the steel bar becomes 1000.24 mm and that of the brass bar 1000.40 mm. The difference in length is 0.16 mm.

On a measured length of 100 mm the difference is

$$^{100}/_{1000} \times 0.16 \text{ mm} = 0.016 \text{ mm or } 16 \, \mu\text{m (16 micrometres)}$$

Measuring at 20°C

In order to eliminate inaccuracies, it has been agreed that all precision measurements will be taken at a temperature of 20°C.

4.c.ii Inaccuracy of measuring equipment

Since measuring instruments have to be manufactured, it is obvious that they are not perfectly accurate and consequently have errors of their own. Measuring instruments should have a greater accuracy than required in the measurements for which they are used.

4.c.iii Reading errors

All measuring equipment with a graduated scale (rules, vernier calipers, micrometers, dial indicators) indicates a size or a size difference which can be read.

The width of the division lines and the imperfections of the human eye may be the cause of reading errors. Reading errors may be present in two places when taking a reading from a rule by the method shown in Fig. **1.4.5**. The possibility of a reading error is reduced to one position in

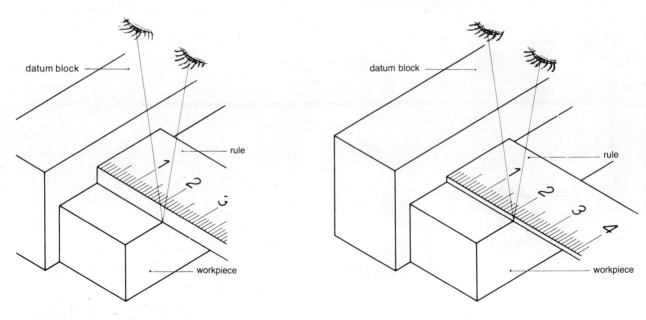

1.4.7 and 8 Rule: the eye should be positioned directly above the reading point

Fig. **1**.4.6. When reading a dimension, the eye should be positioned directly above the point of reading (Figs. **1**.4.7–**1**.4.9).

Figures **1**.4.7 and **1**.4.8 show rules of different thicknesses. If the eye is not positioned directly above the reading point, the possibility of a reading error is greatest with the thickest rule.

The possibility of reading errors is further reduced when using measuring equipment such as a vernier caliper or a micrometer. These are provided with a fixed measuring face and an adjustable measuring face (Figs. **1**.4.10 and **1**.4.11). The moving jaw of a vernier caliper moves directly. When the moving jaw of the vernier caliper in Fig. **1**.4.12 is moved a division of 1 mm on the stock the distance between the fixed jaw and the moving jaw changes similarly by 1 mm.

The movement of the spindle of a micrometer is indirect (Fig. **1**.4.13). By rotating the thimble of a micrometer one division on the sleeve graduation

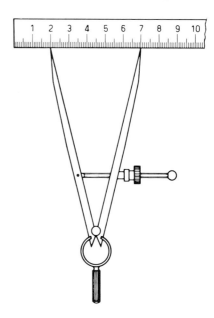

1.4.9 Transferring a dimension with dividers

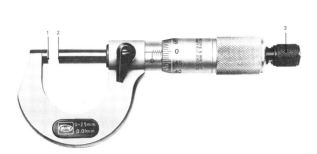

1.4.11 Micrometer measuring faces
1 fixed measuring face (anvil)
2 adjustable measuring face (spindle)
3 ratchet

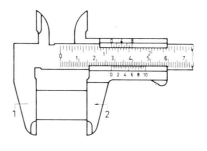

1.4.10 Vernier caliper measuring faces
1 fixed jaw
2 adjustable jaw

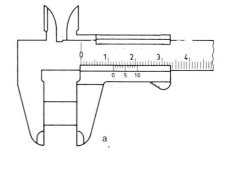

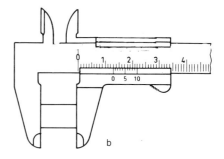

1.4.12 Vernier caliper: slide movement of 1 division on the stock changes jaw separation by 1 mm

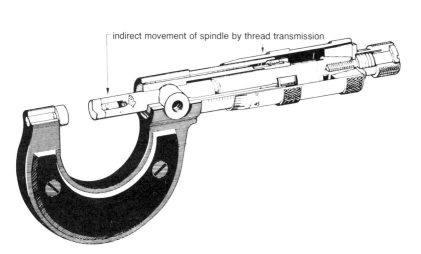

indirect movement of spindle by thread transmission

1.4.13 Construction of a micrometer

(Fig. **1.**4.14), the distance between the anvil and the spindle changes by 0.01 mm. A reading error of 10% on a vernier caliper causes a measuring error of 10% of 1 mm = 0.1 mm. A reading error of 10% on a micrometer thimble causes a measuring error of 10% of 0.01 mm or 0.001 mm if the micrometer has a vernier scale.

4.c.iv Application of force when measuring

As has been indicated, the chance of reading errors is reduced by using a fixed stop and an adjustable stop. However, a further possibility of inaccuracy arises — the **force** applied to the workpiece through the stops. A difference in the force applied, and thus the measuring pressure, may cause inaccuracies of up to 0.05 mm. A constant force is applied by using the ratchet of a micrometer (Fig. **1.**4.11). However, the measuring pressure can be quite different, as illustrated in Fig. **1.**4.15 which refers to three different areas of contact for the same applied force.

4.c.v The skill of the craftsman

Inexpert use of measuring equipment, or inadequate care during measuring, are the cause of most reading errors. Figs. **1.**4.16 and **1.**4.17 show the error that results when the contact faces of the instrument are incorrectly placed on the workpiece.

4.c.vi Cleanliness

Workpieces as well as measuring equipment should be carefully cleaned to ensure accurate measurement and/or checking.

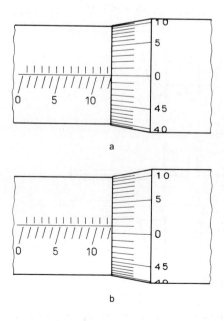

a

b

sleeve division gives a change
of 0.01 mm

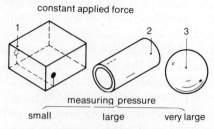

1.4.15   Equal force applied, unequal measuring pressure

1   full surface contact
2   line contact
3   point contact

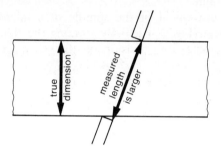

1.4.16a Measuring an outside dimension. A larger reading is obtained owing to incorrect positioning of the instrument

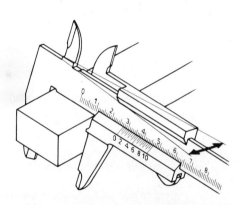

1.4.16b Remedy: rock the vernier caliper gently and feel for play

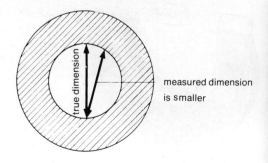

1.4.17   Measuring an internal dimension. A faulty position results in a measured dimension which is too small

4.c.vii Deflection of measuring equipment and workpieces

Deflection of measuring equipment and workpieces may be the cause of substantial measuring errors. (For clarity, an exaggerated illustration is given in Fig. **1.**4.18.) Errors can be prevented by not only measuring the work at different places, but also in different positions. Measuring a thin-walled large cylinder may be carried out more accurately with the cylinder upright (Fig. **1.**4.19) rather than in the horizontal position (Fig. **1.**4.18).

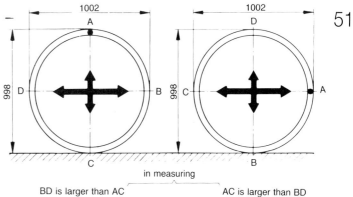

BD is larger than AC   in measuring   AC is larger than BD

1.4.18   Measuring error by deflection

4.d Terminology of measurement related to accuracy

The size entered on the drawing (Fig. **1.**4.20a), is the **specified size**. The high and low limits and the tolerance can be derived from this dimension (Fig. **1.**4.20b). After measurement with precision measuring instruments the **actual size** is determined for the component.

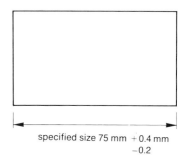

specified size 75 mm +0.4 mm
−0.2

1.4.20a   Specified size

tolerance = 0.6 mm (0.4 mm + 0.2 mm)

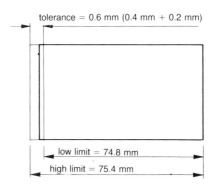

low limit = 74.8 mm
high limit = 75.4 mm

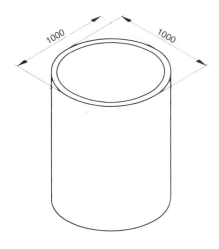

1.4.19   Upright position, causing no deflection

1.4.20b   Limits, tolerance

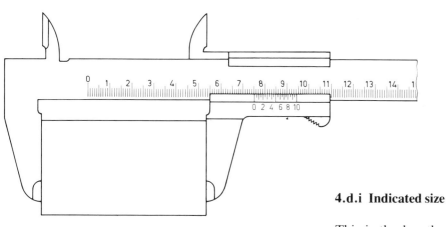

workpiece within tolerance

I.4.21   Indicated size

4.d.i Indicated size

This is the length or angle of the dimension to be measured which is indicated by the measuring equipment (Fig. **1.**4.21).

4.d.ii Reading

This is the length or angle which is read on the measuring instrument. Misreading may cause the observed dimension to deviate substantially from the indicated size.

4.d.iii Mean size

This is the dimension lying halfway between the high and low limit (Fig. **1.4.22**). The mean size is always closer to the true size than one of the limits.

4.d.iv Reading value (graduation)

The graduation of a measuring instrument is the smallest value that can be read from it. This may be 0.02 mm or 0.05 mm for a vernier caliper (Fig. **1.4.12**) or 0.01–0.001 mm for a dial test indicator (Fig. **1.4.23**).

4.d.v Range

The range is determined by the minimum and maximum dimension (length or angle) which can be measured with a measuring instrument.

4.d.vi Instrument accuracy and work tolerance

The accuracy of the measuring instrument depends on the maximum allowable deviation from the indicated size. It is assumed as a general rule that the measuring instrument should have an accuracy

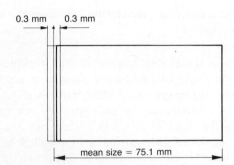

1.4.22 Mean size

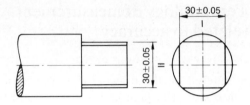

1.4.24

greater than 0.5 to 0.1 of the specified workpiece tolerance in order to ensure that the mean size of the work will lie between the two limits. It is most important, when measuring, that the accuracy of the equipment is known, because the choice of the measuring instrument is directly related to the tolerance specified for the workpiece.

In summary, the accuracy of measurement is the addition of all deviations, e.g. the difference in measuring pressure and reading errors, which are likely to occur when measuring is performed proficiently.

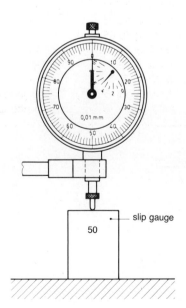
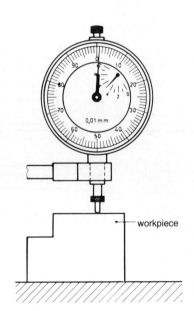

1.4.23 The dial test indicator is calibrated in 0.01 mm

4.e Graduation and accuracy

It is emphasised that there is a difference between
- the graduation value and
- the accuracy.

A graduation to read 0.01 mm does not necessarily imply an accuracy of 0.01 mm. Generally, an instrument has a greater accuracy than the graduation. The greater the accuracy the less chance there is for errors.

Example 1 A round shaft has to have a square at one end (Fig. **1**.4.24). The vernier caliper to be used has a vernier scale graduated to read $\frac{1}{20}$ mm. The instrument accuracy is ± 0.05 mm. The readings taken are respectively
- 29.95 mm
- 30.05 mm

The actual sizes may be
- **from** 29.95 mm $-$ 0.05 mm $=$ 29.90 mm
  to 29.95 mm $+$ 0.05 mm $=$ 30.00 mm
- **from** 30.05 mm $-$ 0.05 mm $=$ 30.00 mm
  to 30.05 mm $+$ 0.05 mm $=$ 30.10 mm

In the most favourable case the actual sizes may be 30.00 mm, but in the most unfavourable case 29.90 mm and 30.10 mm, which are both beyond limits.

Example 2 Ten shafts have to be turned to a diameter of 30 mm. The vernier scale is graduated to read $\frac{1}{20}$ or 0.05 mm. The vernier caliper accuracy is ± 0.05 mm. Table **1**.4.4 gives a survey of the correct readings taken and the chance of rejection.

The accuracy of the measuring instruments in *Examples 1* and *2* is inadequate in relation to the specified workpiece tolerance. An approximation to the mean size could have been achieved more effectively by using a micrometer which has greater precision. There is less chance of deviation with such an instrument.

Table 1.4.4 Correctly read size and the chance of rejection in *Example 2*

Specified size (mm)	Limits (mm)	Reading (mm)	Actual size may be (mm)	Chance of rejection
30 ± 0.1	29.90–30.10	29.90	29.85 or 29.95	× too small
30 ± 0.1	29.90–30.10	29.90	29.85 or 29.95	× too small
30 ± 0.1	29.90–30.10	29.95	29.90 or 30.00	—
30 ± 0.1	29.90–30.10	30.0	29.95 or 30.05	—
30 ± 0.1	29.90–30.10	30.0	29.95 or 30.05	—
30 ± 0.1	29.90–30.10	30.0	29.95 or 30.05	—
30 ± 0.1	29.90–30.10	30.0	29.95 or 30.05	—
30 ± 0.1	29.90–30.10	30.05	30.00 or 30.10	—
30 ± 0.1	29.90–30.10	30.10	30.05 or 30.15	× too large
30 ± 0.1	29.90–30.10	30.10	30.05 or 30.15	× too large

5 Determination of position, level and size

Types of equipment

5.a Surface tables and surface plates

5.a.i Sizes

The main sizes of surface tables and plates are specified by the length and width of the top surface.

Table 1.5.1 Cast iron surface tables (Fig. 1.5.1) and surface plates (Fig. 1.5.2)

Length (mm)	Width (mm)	Mass (kg)
150–3000	150–1500	5–2300

Table 1.5.2 Granite surface tables and surface plates (Fig. 1.5.3)

Length (mm)	Width (mm)	Thickness (mm)	Mass (kg)
200–2500	200–1500	60–260	8–2500

5.a.ii Application

Surface tables and plates are used solely for marking out and for the checking of workpieces. Figs. 1.5.4–1.5.8 show respectively

1.5.2 Cast iron surface plate

- marking out parallel lines with a scribing block and square lines with a try-square (Fig. 1.5.4)
- checking parallelism with a dial test indicator (Fig. 1.5.5)
- marking out and measuring with slip gauges (Fig. 1.5.6).

5.a.iii Care of surface tables and surface plates

Cast iron surface tables and plates require more maintenance than those made from granite. Both types have to be cleaned after use. Cast iron surface tables and plates need to be oiled with acid-free oil. These surface tables and plates should not be used as a surface on which to store materials or tools. When not in use, they should be protected by a wooden cover.

1.5.1 Cast iron surface table

1.5.3 Granite surface table

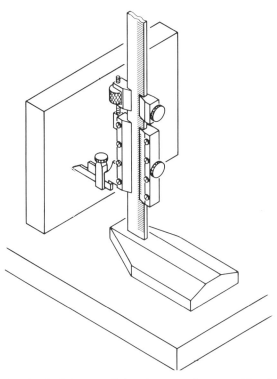

1.5.4   Scribing parallel and square lines, working from a surface plate

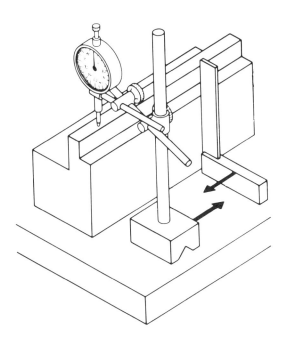

1.5.5   Checking parallelism on a surface plate, using a dial test indicator

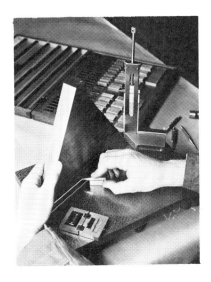

1.5.6   Marking out on a surface plate, using a scriber mounted on a slip gauge with a protector slip

1.5.7   Checking parallelism on a surface plate and using slip gauges

1.5.8   Checking of lengths, by using slip gauges on a surface plate. The slip gauges are clamped in a fixture

5.b Vee-blocks

5.b.i Sizes

Table 1.5.3 Vee-block (Fig. **1.**5.9)

Length (mm)	Width (mm)	Height (mm)
50	40	40
75	60	60

1.5.9   Vee-block and clamp

Table 1.5.4 Vee-block (Fig. 1.5.10)

Length (mm)	Width (mm)
100	40
150	50
200	70
250	85
300	100

5.b.ii Application

Vee-blocks are used on a surface plate to support and hold cylindrical workpieces. For long workpieces they are made in matched pairs.

5.b.iii Care of vee-blocks

Vee-blocks must be handled with care to prevent damage. A damaged base, for instance, may cause errors in marking out.

1.5.10   Vee-block

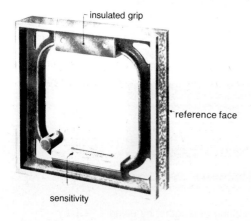

insulated grip

reference face

sensitivity

1.5.11   Square block level

1.5.12   Block level

5.c Spirit levels

5.c.i Sizes

The main dimensions are expressed as the length and width (*Figs*. 1.5.11 and 1.5.12)

Table 1.5.5 Square block level (Fig. 1.5.11)

Length (mm)	Width (mm)
150	30
200	40
250	45

Table 1.5.6 Block level (Fig. 1.5.12)

Length (mm)	Width (mm)
160	43
200	50
250	56
300	63

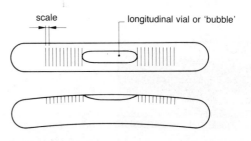

scale

longitudinal vial or 'bubble'

1.5.13   Vial

1.5.14   Level with adjusting screw:
        sensitivity 0.02 mm/m

5.c.ii Nomenclature

The different parts of a level are given in Fig. **1**.5.11. The square block level shown has four reference planes. The lower and right-hand surface are vee-shaped. Fig. **1**.5.13 shows a vial (the graduated glass liquid container) in detail. Fig. **1**.5.14 shows a level with an adjusting screw to set the level to the correct position for checking inclined surfaces.

5.c.iii Sensitivity

More important than either the length or width of a level, is the sensitivity. It is the amount of deviation from the horizontal or vertical plane, expressed in millimetres per metre for each vial division (Fig. **1**.5.13).

Example The measured length is 5 m. The level sensitivity is 0.02 mm/m. The deflection of the vial is 3 divisions. The deviation is

$$3 \times 5 \text{ m} \times 0.02 \text{ mm/m} = 0.3 \text{ mm}$$

5.c.iv Application

Levels are used for levelling machine tools, machine parts and fabrications (Figs. **1**.5.15–5.5.18).

5.c.v Care of levels

Precision spirit levels are instruments which have to be handled with great care. They must be cleaned after use, the reference surfaces oiled, and then put away in their cases.

5.d Straight edges

5.c.i Sizes

Straight edges, as shown in Figs. **1**.5.19 and **1**.5.20, are available in the following sizes

Table **1**.5.7 Master straight edges (Fig. **1**.5.19)

Length (mm)	Width (mm)	Thickness (mm)
From 500	50	10
To 5000	140	20

Table **1**.5.8 Workshop straight edges (Fig. **1**.5.20)

Length (mm)	Width (mm)	Thickness (mm)
From 500	30	6
To 5000	120	8

Precision straight edges (Fig. **1**.5.21) are manufactured in lengths from 70 mm to 1000 mm.

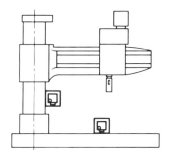 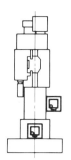 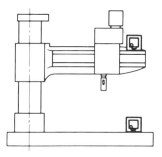 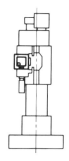

1.5.15 to 18   Levelling and checking a radial drilling machine, using a square block level

1.5.19   Master straight edge

1.5.20   Workshop straight edge

1.5.21   Precision straight edge

5.d.ii Application

Master and precision straight edges should not be used for purposes other than checking flatness. Workshop straight edges may be used for both checking and marking off. Figs. **1.5.22–1.5.26** show examples of applications of straight edges.

5.d.iii Care of straight edges

Straight edges in general should be cleaned after use and, where necessary, oiled. Precision straight edges should be put away in their cases after use.

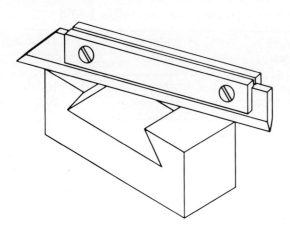

1.5.22 Checking for flatness of a dovetail slide

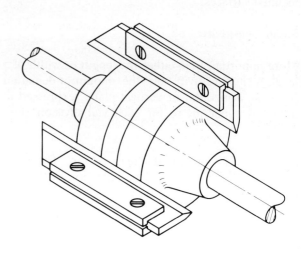

1.5.23 Checking for alignment of a coupling

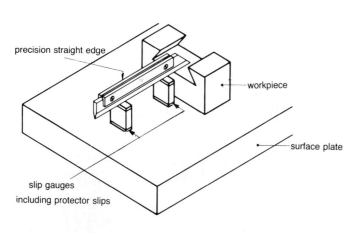

1.5.24 Measuring, using slip gauges and a precision straight edge

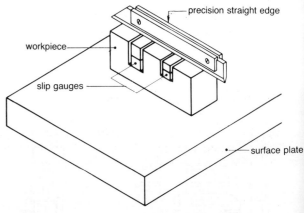

1.5.25 Measuring, using slip gauges and a precision straight edge

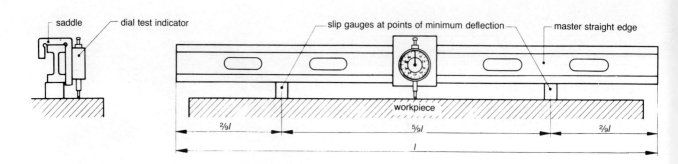

1.5.26 Checking, using a master straight edge and a dial test indicator

5.e Feeler gauges

5.e.i Sizes

The principal dimensions are the blade thickness and blade length (Fig. **1**.5.27).
Common dimensions are
- blade length 100 mm
- thickness 0.03 mm, up to 1 mm inclusive, in increments of 0.01 mm
- thickness 0.1 mm up to 2 mm inclusive, in increments of 0.05 mm

5.e.ii Long blade feeler gauges

The dimensions of long blade feeler gauges as shown in Fig. **1**.5.28 are length, and thickness, from 0.05 mm to 1 mm, in increments of 0.05 mm.

5.e.iii Application

Feeler gauges are used to measure the clearance between two components, e.g. a shaft in its bearing, or the clearance of an engine valve (internal combustion). The blades may be used singly or in combination.

5.e.iv Care of feeler gauges

Feeler gauges should be cleaned after use, and oiled. Even slight corrosion may alter the size.

5.f Cylindrical pin gauges

Fig. **1**.5.29 shows a set of cylindrical pin gauges. Their diameters range from 0.45–3.0 mm inclusive.

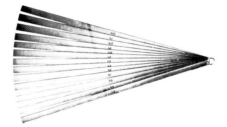

1.5.28   Long blade feeler gauges

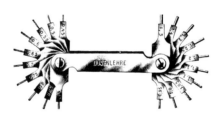

1.5.29   Cylindrical pin gauges

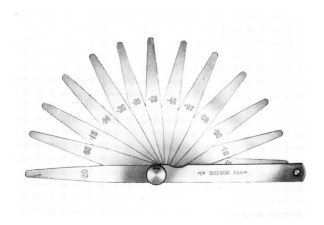

1.5.27   Feeler gauges

5.g Taper gauges

5.g.i For small holes

These gauges may be used to measure holes up to 3 mm diameter (Fig. **1**.5.30). Readings are accurate to 0.1 mm.

5.g.ii For medium sized holes

These gauges may be used to measure holes of 1–6 mm, 4–15 mm and 15–30 mm diameter. Reading is accurate to 0.1 mm (Fig. **1**.5.31).

General rules for accurate measuring

If observed, the following general rules should ensure measurement to the accuracy required for any given situation and reduce the chances of rejections.

- Ascertain the tolerance applicable to the dimension which has to be measured, and select measuring equipment with a permissible manufactured tolerance not exceeding 0.5–0.1 mm of that tolerance.
- Be satisfied that the measuring equipment chosen is fully serviceable, e.g. you have inspected it for cleanliness and damage; you know that it is subject to regular checks for accuracy.
- Ensure that the equipment and workpiece are at similar temperatures and as near to the ideal of 20°C as is practicable.
- Take measurements as close to the mean size as possible and follow the rules of procedure for accuracy appropriate to the specific operation, e.g. when reading a rule, have the eye directly above the mark.

1.5.30 Taper gauge for small holes

1.5.31 Taper gauge for medium sized holes

Part 2

1 Marking out

1.a Purpose

The purpose of marking out is to transfer **to** workpieces (Figs. **2**.1.1 and **2**.1.2) **from** drawings (Fig. **2**.1.3) the lines needed for working. Centre marks are made on the scribed lines if needed (Fig. **2**.1.2).

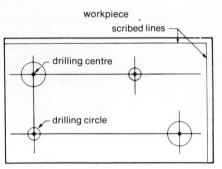

2.1.2 Front view of the marked out workpiece in Fig. **2**.1.1

1.b Methods

1.b.i Marking out lines

There are two main methods of marking out lines on workpieces.
- Scribing with hard-pointed tools such as a scriber (Fig. **2**.1.4), a vernier height gauge with scriber (Fig. **2**.1.5) or dividers (Fig. **2**.1.6). This method is used on unfinished or bright material, and work which has previously been coated with a marking medium (engineer's blue, copper sulphate solution, chalk, etc.).
- Marking with a scriber made of a softer material, e.g. a brass scriber on sheet steel or a pencil on sheet aluminium.

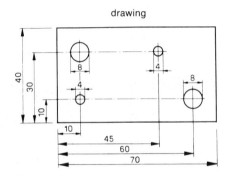

2.1.3 Drawing for marking out the workpiece in Fig. **2**.1.1

1.b.ii Centre marking

Centre marks are made with a centre punch (Fig. **2**.1.7) and hammer, or with an automatic centre punch (Fig. **2**.1.8). There are also other types of punch — one, for instance, has a triangular point.

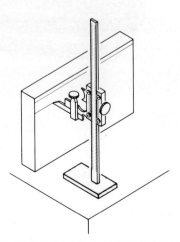

2.1.1 Marking out using a vernier height gauge

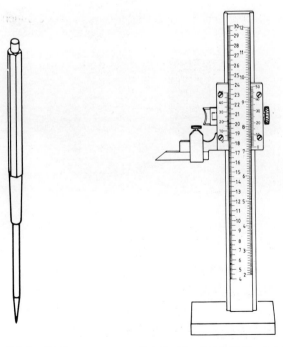

2.1.4 Scriber

2.1.5 Vernier height gauge with s

2.1.6   Dividers

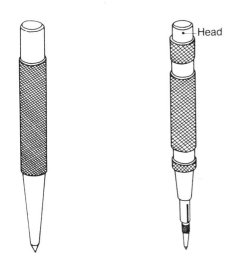

2.1.7   Centre punch   **2.**1.8   Automatic centre punch

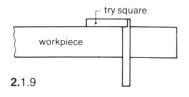

2.1.9

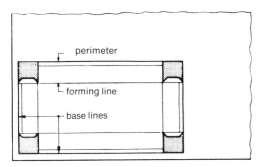

2.1.10

1.b.iii Applications

Some of the applications of the marking out process are given below.

- To mark cutting lines (Fig. **2.**1.9) on bar stock and sections.
- To mark perimeter lines, bending and forming lines on sheet materials (Fig. **2.**1.10).
- To indicate the shape of the workpiece on cut-off material (Fig. **2.**1.11).
- To indicate round or rectangular holes, slots and shapings on partially finished workpieces (Fig. **2.**1.12).
- To mark mitres on sections (Fig. **2.**1.13).
- To mark centres on the ends of shafts (Fig. **2.**1.14).

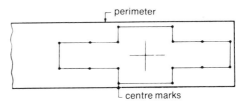

2.1.11

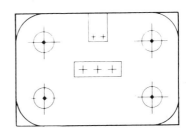

2.1.12

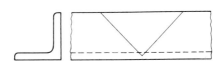

2.1.13

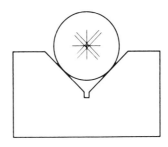

2.1.14

1.c Scribed lines and drawn lines

1.c.i Scribed lines

When the scriber is harder than the workpiece material, the point will scratch (scribe) a groove into the material (Fig. 2.1.15a and b). It does the same to a marking medium (Fig. 2.1.16a and b).

1.c.ii Drawn lines

When the scriber is softer than the workpiece material, the point leaves a drawn (surface) line.

For this reason there must be a distinct colour difference between the scriber and the workpiece material, e.g. a brass scriber used on steel, a soft pencil used on aluminium (Fig. 2.1.17a and b) or a chalk line (Fig. 2.1.18a and b).

1.c.iii Characteristics of scribed and drawn lines

- Scribed lines are thin, sharply defined and cannot be removed without further working (Fig. 2.1.19a and b).
- Drawn lines are thicker and less sharply defined. They lie on the surface and are usually easy to remove (Fig. 2.1.20a and b).

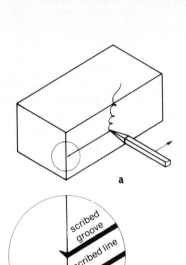

2.1.15 Scribed line

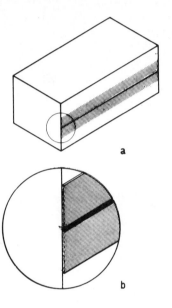

2.1.16 Line scribed into a marking medium

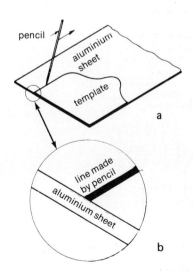

2.1.17 Drawn line

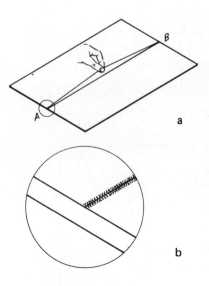

2.1.18 Chalk line

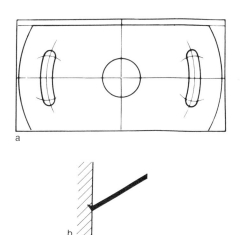

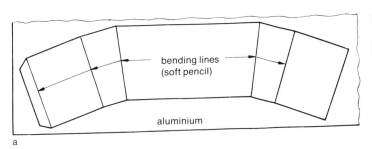

bending lines
(soft pencil)

aluminium

2.1.19   Scribed line

2.1.20   Drawn line

1.d Centre punching

1.d.i Method

Support the hand holding the centre punch. Holding the punch at an angle so that the point and the intersection of scribed lines remain visible,

bring the punch upright and strike it (Fig. **2.1**.21). The shape and depth of the indentation must provide for the accurate use of dividers, or the entry of drills, so the centre marks for correctly ground points of dividers should not be made too large (Figs. **2.1**.22 and **2.1**.23).

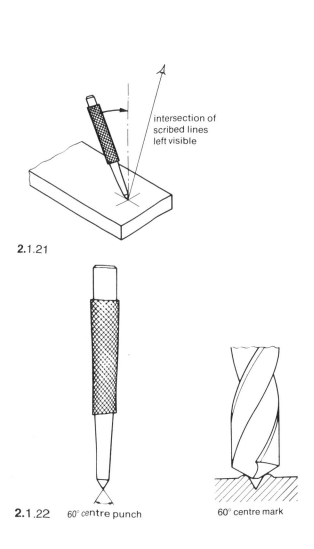

intersection of
scribed lines
left visible

2.1.21

2.1.22   60° centre punch

60° centre mark

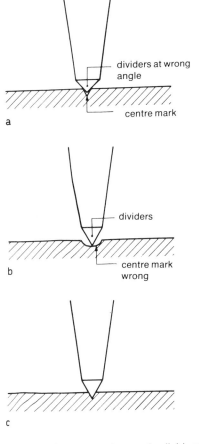

dividers at wrong
angle

centre mark

a

dividers

centre mark
wrong

b

c

2.1.23a Correct centre mark; dividers point wrong

2.1.23b Incorrect centre mark

2.1.23c Centre mark and dividers both correct

When centre marks are punched for drilling, the cutting angle and the diameter of the dead centre of the drill have to be considered (Fig. **2**.1.24). In general, a 60° centre punch (Fig. **2**.1.22) can be used to make centre marks for drilling. The drill and the centre mark have to be carefully aligned. When the pitch of holes is less important, a 120° centre punch can be used and the drill aligned directly (Fig. **2**.1.25).

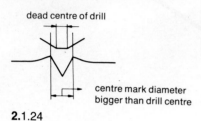

dead centre of drill

centre mark diameter bigger than drill centre

2.1.24

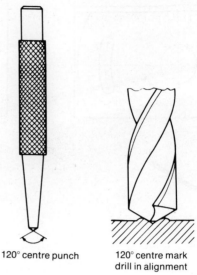

120° centre punch   120° centre mark drill in alignment

2.1.25 Centre punch (120°) for aligning a drill for non-precision drilling. (It is not a marking-out tool)

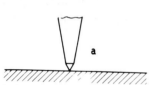

a

before striking with the hammer

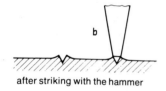

b

after striking with the hammer

2.1.26 Marking the centre with a centre punch

The material around the centre mark will always be slightly deformed because the point of the centre punch is driven into the material (Fig. **2**.1.26a and b). This may give rise to hair-line cracks or deformation of the opposite face of thin sheet material.

Where necessary, to provide a witness of a scribed line after removing metal from a workpiece, e.g. machining, it is usual to centre punch the lines lightly before commencing the operation.

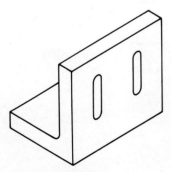

2.1.27 Angle plate

1.e Marking out equipment

Marking out equipment may be divided into three main types
- those which support the workpiece or marking instrument
- those from which dimensions are transferred
- those with which or along which scribed lines are made.

1.f Form and use of specific items

1.f.i Support or datum surfaces

A datum surface provides a consistent surface from which measurements may be taken when marking out. Datum surfaces include the angle plate (Fig. **2.**1.27), the adjustable angle plate (Fig. **2.**1.28) and the surface table (Fig. **2.**1.29). Support and datum surfaces include rollers (Fig. **2.**1.30), parallels (Fig. **2.**1.31), parallel blocks (Fig. **2.**1.32) and vee-blocks (Figs. **2.**1.33 and **2.**1.34).

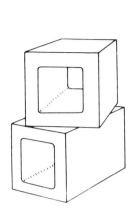
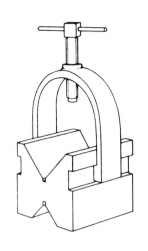

2.1.32   Parallel blocks   **2.**1.33   Vee-block and clamp

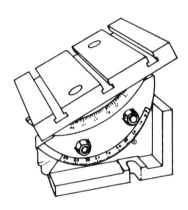

2.1.28   Adjustable angle plate

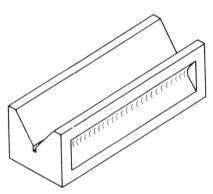

2.1.34   Vee-block

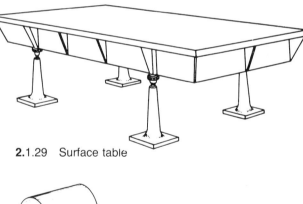

2.1.29   Surface table

2.1.30   Rollers

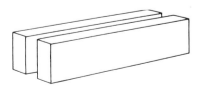

2.1.31   Parallels

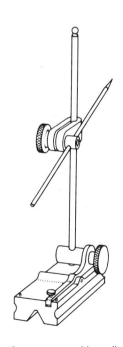

2.1.35   Surface gauge with scriber

1.f.ii Gauges

Surface gauges

Surface gauges (Fig. **2.**1.35) are placed on a surface table (Fig. **2.**1.29) and are used when lines parallel to it are to be scribed on the workpiece. A steel rule placed against an angle plate is sometimes used, with a surface gauge to set the scriber at the required height.

Height gauge

The dimension can be read directly from the height gauge shown in Fig. **2.**1.36. (See also Fig. **2.**1.60.)

Marking out machine

The marking out machine shown in Fig. **2.**1.37 has a telescopic horizontal arm which is raised or lowered on a vertical pillar. The whole machine can be moved across the surface table. The end of the horizontal arm has a swivel head to hold marking out instruments. If required, the machine can be mounted on a rotary table to increase its versatility.

Slip gauges

These are hardened steel blocks with opposing faces lapped flat and parallel to a definite size, within extremely tight tolerances. They are usually made in sets to one of four grades in the relevant

2.1.37   Using a marking out machine

2.1.36   Vernier height gauge

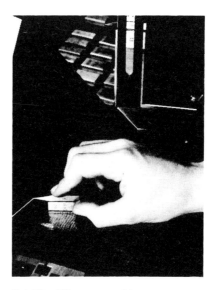

2.1.38   Slip gauge with tungsten carbide protector slip

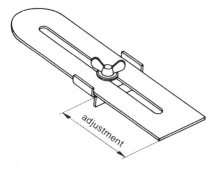

2.1.39   Adjustable scribing gauge for parallel lines

BSI specifications. Slip gauges can be used when very accurate marking out is required (Fig. **2.**1.38). In use extreme cleanliness is essential and a tungsten carbide protector slip should be interposed between the surface table and the base block to protect it from wear.

2.1.40   Odd-leg calipers

Adjustable scribing gauge
The adjustable scribing gauge (Fig. **2.**1.39) is used mainly for scribing lines parallel to the side of a workpiece.

Odd-leg calipers
These calipers (also called jennies, single-leg calipers or hermaphrodite calipers) are also used to scribe lines parallel to an edge (Fig. **2.**1.40).

1.f.iii Marking equipment

Box square
A box square (Fig. **2.**1.41) is used to scribe lines along a shaft.

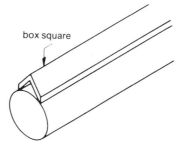

2.1.41   Box square

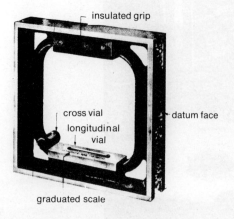

2.1.42   Square block level

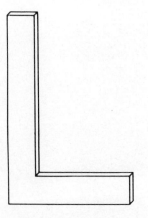

2.1.43   Flat try-square

Surface plate

The top surface of a surface table should preferably be perfectly horizontal. The square block level (Fig. 2.1.42) is used to set it.

Block level

The square block level can also be used to check that surfaces of workpieces are placed truly horizontal or truly vertical.

Equipment for right angles and perpendiculars

The instruments shown in Figs. 2.1.43–2.1.46 are used to set right angles and perpendiculars on the surface table. Those shown in Figs. 2.1.43 and 2.1.45 may also be used to scribe lines at right angles to the edge of a workpiece.

Mitre square

Mitre squares, sometimes called bevels (Fig. 2.1.47), are used similarly for setting or scribing mitres.

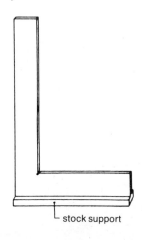

2.1.44   Flat try-square with stock support

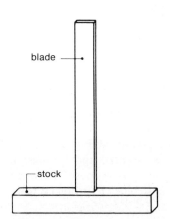

2.1.46   T-square with stock and blade

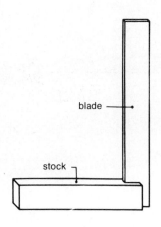

2.1.45   Square with stock and blade

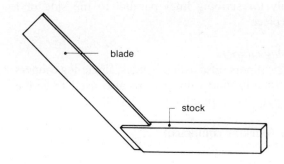

2.1.47   Mitre square

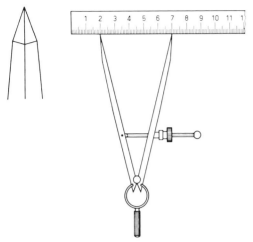

2.1.48   Dividers

2.1.53   Measuring tape

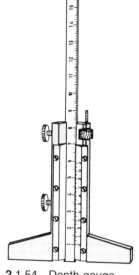

2.1.54   Depth gauge

Dividers

Dividers are used for scribing circles or marking off equal distances (pitch). The pivot of the dividers should allow the legs to move freely, smoothly and true so that the dimension to which they were adjusted is kept constant. The points must be ground, as shown in Fig. **2.**1.48.

Trammels

Trammels (Fig. **2.**1.49) are used to scribe large diameter circles and arcs. Some models have a fine adjusting device at one end.

Centre square

The centre square (Fig. **2.**1.50) is used to find the centre of the end of a shaft.

Linear measuring equipment

The equipment shown in Figs. **2.**1.51–**2.**1.54 is used for setting out and transferring linear measurements.

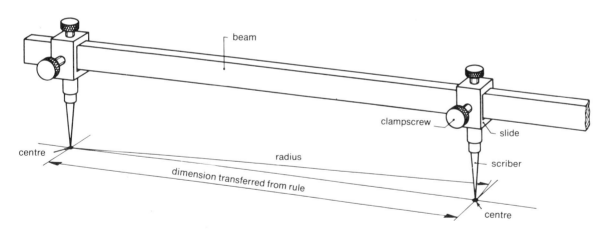

2.1.49   Trammels

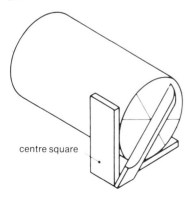

2.1.50   Centre square

2.1.51   Steel rule

2.1.52   Wooden rule

1.g Materials used for marking out equipment

Equipment	Material
Angle plate	cast iron
Vee-block	cast iron
Surface plate	cast iron or granite
Surface and height gauges	
— base	usually cast iron
— rest	high quality steel
Parallels	hardened steel
Scribers, centre punches and points of dividers	hardened tool steel (carbide tips are used increasingly on scribing tools)
Squares, rules, tapes	high quality steel
Soft lead pencils	mixture of graphite and clay

1.h Marking out lines and circles

Using the surface plate
Much marking out of lines involves the use of the surface plate, as shown in Fig. **2.1.55**. However, there are other methods used.

Using a rule
A rule used for marking off dimensions should be placed against a datum block, as shown in Fig. **2.1.56**.

1.h.i Straight lines

If one side of a small workpiece is flat and square, one dimension mark will be enough for each line which has to be scribed (Fig. **2.1.57**). Two marks, as far apart as possible, will be needed for each line on a large workpiece (Fig. **2.1.58**).

1.h.ii Circles and arcs

The dividers are set to the correct radius, as shown in Fig. **2.1.59**. The points are placed right against the graduation marks to avoid inaccuracy.

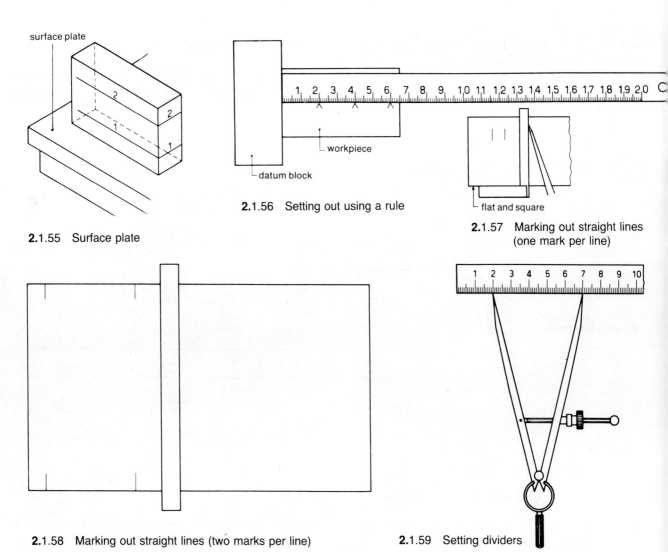

2.1.55 Surface plate

2.1.56 Setting out using a rule

2.1.57 Marking out straight lines (one mark per line)

2.1.58 Marking out straight lines (two marks per line)

2.1.59 Setting dividers

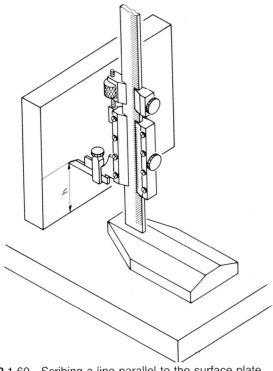

2.1.60   Scribing a line parallel to the surface plate using a height gauge

1.h.iii Lines parallel and perpendicular to the surface plate

Figure **2.1.60** shows how a height gauge is used for scribing lines parallel to the surface plate. Fig. **2.1.61** shows a square and scriber being used for scribing lines perpendicular to the surface plate. Thin sheet workpieces are placed against an angle plate, or clamped to it, before marking out.

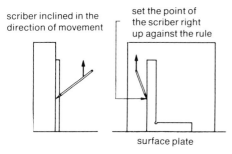

2.1.61   Scribing a line perpendicular to the surface plate using a square and a scriber

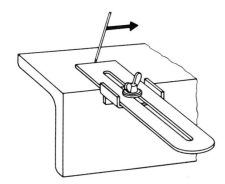

2.1.62   Scribing a parallel line on a section using an adjustable scribing gauge

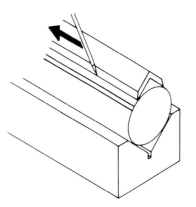

2.1.63   Scribing a parallel line on a shaft using a box square

1.h.iv Parallel lines on angled sections

An adjustable scribing gauge (Fig. **2.1.62**) is used to scribe lines parallel to the edge of an angled section.

1.h.v Parallel lines on shafts

A box square (Fig. **2.1.63**) or a surface plate and height gauge (Fig. **2.1.64**) may be used to scribe parallel lines along shafts.

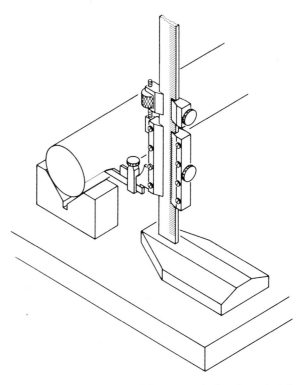

2.1.64   Scribing a parallel line on a shaft using a height gauge on a surface plate

1.i Advantages and limitations of marking out

1.i.i Advantages

Within certain degrees of accuracy, marking out shows
- the boundaries of the workpiece and the amount of material to be removed
- the line of a bend
- the position of a hole
- the size of a hole
- the possibility, or otherwise, of using the material for that size of workpiece.

1.i.ii Limitations

Scribed lines may cause tears in the surface of the workpiece material (Fig. **2**.1.65). Lines drawn with a brass scriber or soft leaded pencil can be too thick or indistinct. Some operations, e.g. precision drilling, require a greater degree of accuracy.

1.j. Marking out from various types of datum

1.j.i Marking out from a single datum point

This method is used for scribing circles on discs, rings and other circular objects. It is illustrated in Figs. **2**.1.66, **2**.1.67 and **2**.1.68 and the sequence of operations is numbered.
The centre may be located by using
- a centre square (Fig. **2**.1.50)
- odd-leg calipers (Fig. **2**.1.40)
- the procedure illustrated in Fig. **2**.1.69 if the diameter of the disc is large.

A metal strip attached to a piece of wood can be used when the centre of a ring has to be located.

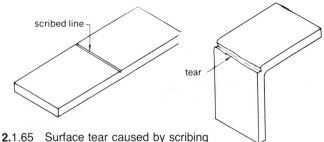

2.1.65   Surface tear caused by scribing

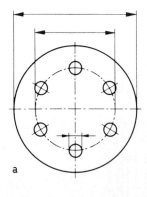

a

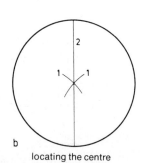

b
locating the centre

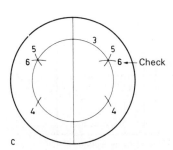

c

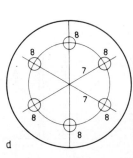

d

2.1.66   Marking out a flange from a single datum point

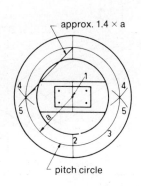

2.1.67   Marking out a ring

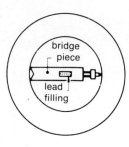

2.1.68   Auxiliary tool for marking out a ring

2.1.69   Finding the centre of a large disc

1.j.ii Marking out from a single datum point and a single datum line

This method of marking out is used when workpieces are symmetrical. In Fig. **2.**1.70 the marks on either side of the line numbered 1 are identical. The rest of the marking out is done in the numbered sequences. The straight lines must be drawn along a straight edge.

When the centre point of an arc lies outside the workpiece, another piece of the same height or thickness has to be used (Fig. **2.**1.71). Both the workpiece and the other piece have to be clamped down.

1.j.iii Marking out from one or more datum faces

Marking out of a workpiece from one or more datum faces is done on the surface plate. The datum faces of the workpiece shown in Fig. **2.**1.72 are easy to identify because the dimensions are taken from them. The datum face rests on the surface plate during marking out, and if the workpiece is in thin sheet, it is supported by (or clamped to) an angle plate.

The sequence for marking out is indicated by the numbers in Figs. **2.**1.73 and **2.**1.74. The vertical lines numbered 2 are drawn with a square. If a workpiece is to be finished on all sides (Fig. **2.**1.75a), it also has to be marked out on the surface plate and the face from which dimensions are taken has to rest on the surface plate. Fig. **2.**1.75b, c and d show the sequence in which marking out may be

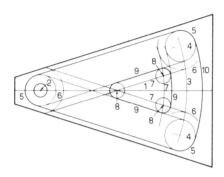

2.1.70   Marking out a symmetrical workpiece

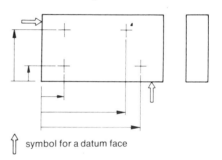

symbol for a datum face

2.1.72   Drawing of a rectangular workpiece with datum faces

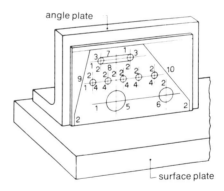

2.1.73   Marking out a workpiece in thin sheet held against an angle plate

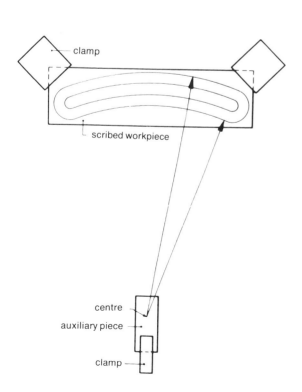

2.1.71   Marking out when the centre point of an arc lies outside the workpiece

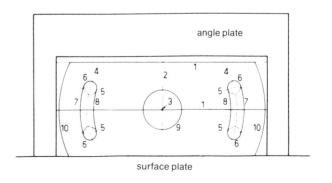

2.1.74   Marking out a workpiece in thin sheet held against an angle plate

done. Other sequences are acceptable, but lines with the same number must always be marked out consecutively, at the same setting of the workpiece.

Note Fig. **2.**1.76a shows the unacceptable accumulation of tolerances resulting from what is known as chain dimensioning. This accumulation is avoidable by progressive dimensioning from a common datum, as shown in Fig. **2.**1.76b.

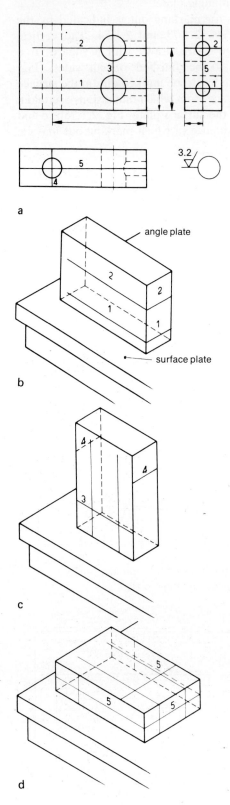

a

b

c

d

2.1.75 Marking out a workpiece using a surface plate as a datum

1.j.iv Marking out using the centre of a hole as datum

In Fig. **2.**1.77, the dimensions of the workpiece are taken from the centre of the hole. The hole and the four surfaces have been pre-machined on a lathe. The workpiece is mounted on a stud fastened to a plate and supported, if necessary, by a jack. The workpiece may be rotated on the stud so that the straight lines parallel to the surface plate may be scribed with the aid of a height gauge.

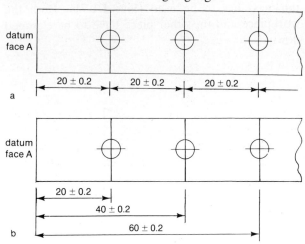

datum face A

20 ± 0.2   20 ± 0.2   20 ± 0.2

a

datum face A

20 ± 0.2

40 ± 0.2

60 ± 0.2

b

2.1.76 Examples of
a 'chain dimensioning: maximum tolerance from datum face A ± 0.6

b 'progressive dimensioning': by which maximum tolerance from datum face A is held to ± 0.2
(all dimensions are in millimetres)

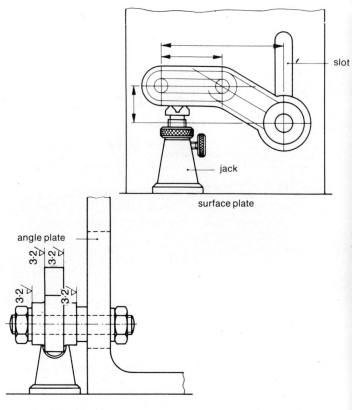

2.1.77 Marking out a workpiece mounted on a stud in an angle plate

1.k Using coordinates

Marking out by hand is not always sufficiently accurate. A specially equipped drilling machine can be more accurate and save time.

To understand how the machine works, look at Fig. **2.**1.78. Imagine that a pencil is placed in the drill chuck and that a piece of paper is glued to the table. When the table moves, a visible line is made. If the table is moved lengthways the x-axis is drawn, and if it moves crossways the y-axis is drawn. Where the two axes cross is the datum point or point of origin (O). The z-axis is a perpendicular from this point, and is the centre line of the pencil, so any movement of the pencil is a change of height (z-axis) while the table moves lengthways (x-axis) or crossways (y-axis).

Accuracies of 0.01 mm can be achieved. This degree of accuracy is the reason why the dimensions on the x- and y-axes are given in hundredths of millimetres in the drilling operations sheet for a flange (Fig. **2.**1.79a) in which six holes have to be drilled. In a real operation, of course, it

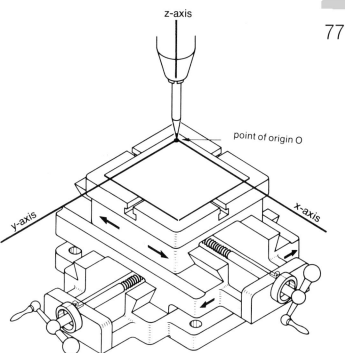

2.1.78 Principle of marking out using coordinates

x	y	Hole no.	Sequence of operations		x	y
(mm) 000·000	(mm) 000·00		mounting	⊗	(mm) 400·00	(mm) 300·00
337·50	191·75	1	drill 8	A	400·00	120·00
462·50	–	2	drill 18	B	220·00	300·00
525·00	300·00	3	drill 18			
462·50	408·25	4	drill 18			
337·50	–	5	drill 18			
275·00	300·00	6	drill 18			
000·00	000·00		removing			

Drilling operations sheet for a flange

a

2.1.79 Marking out a flange using coordinates:
(a) the drilling operations sheet
(b) and (c) the x- and y-values taken from different datum points (O_1 and O_2)

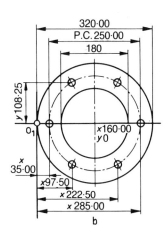

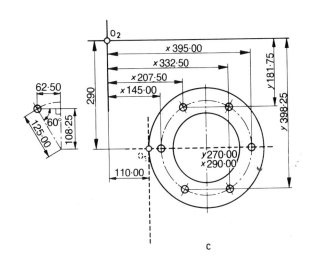

is a drill (not a pencil) in the chuck and the workpiece (not a piece of paper) lies on the table; there is no visible line made — but the principle is the same.

Look at the instructions on the drilling operations sheet in Fig. **2.**1.79a. First work out the distances from the datum point 'O' along the x- and y-axes (called the x and y values) of the centre of the flange. Then work out the x- and y-values of the stops A and

B and finally of each hole. Fig. **2.**1.79b and c shows the same flange, this time marked out from two datum points O_1 and O_2.

Specialist books on drilling and coordinates take this subject further.

Note Look carefully at the working drawings and the location of the holes. You can see that if the tolerances are less than 0.2 mm, the marking out will not be sufficiently accurate.

1.1 Care of marking out equipment

'Good tools halve the job.' It goes without saying that marking out equipment should be kept in good serviceable condition.

Points to remember include

- use the instrument only for the purpose for which it was designed. A surface plate should never, for example, be used as an anvil.
- careful cleaning after use of a surface plate; use a cleaning agent and then apply a thin film of non-acid oil and cover it with a felt-lined wooden cover to prevent accidental damage
- careful cleaning and oiling of all marking out equipment after use
- maintaining correct angles on centre punches (Figs. **2.**1.22 and **2.**1.25) and dividers and scribers (Fig. **2.**1.80b); note particularly that scribers used with a vernier height gauge (Fig. **2.**1.81) should never be ground on the underside.

1.m Productivity in marking out

- Before beginning marking out, check the list of equipment you will need and make sure all items are in good condition.
- Place the workpiece and yourself so that you have proper light for your work.
- Put the equipment within easy reach, with items you need most often placed nearest to you (Figs. **2.**1.82 and **2.**1.83). Each item should have its own place where it will be safe from accidental damage.
- All lines and centre marks which are identical on several workpieces should be marked out together.
- Mark out only what is necessary for accurate working.
- Think whether a template would be justified and whether the first workpiece could be used as a template.
- Use marking out instruments from which dimensions can be read directly.
- Use sharp instruments.
- For better results on stainless steel, use a centre punch with a triangular-shaped point.

1.m.i Safe movement of marking out equipment

To avoid damage, all marking out equipment should be moved carefully. Items such as scribers and dividers should not be carried in the pocket. Their points should be protected by corks (Fig. **2.**1.84), to prevent damage to or by them.

2.1.80a Point angle of a pin punch

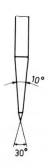

2.1.80b Point angles of a scriber

2.1.81 Correct profile of a scriber used with a vernier height gauge

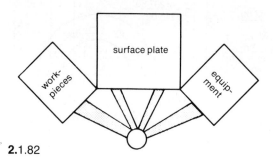

surface plate

work-pieces

equip-ment

2.1.82

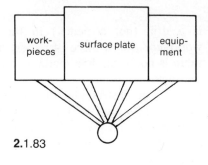

| work-pieces | surface plate | equip-ment |

2.1.83

1.m.ii Safe movement of workpieces

To move heavy objects like large camshafts and frames (Fig. **2.**1.85) on to surface tables, travelling cranes are required. Other objects too heavy for safe manual lifting require hand controlled hoists, or wheel stands (Fig. **2.**1.86).

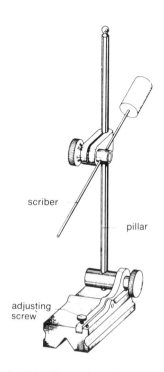

2.1.84 Protection of a surface gauge scriber

2.1.85 Moving a large workpiece

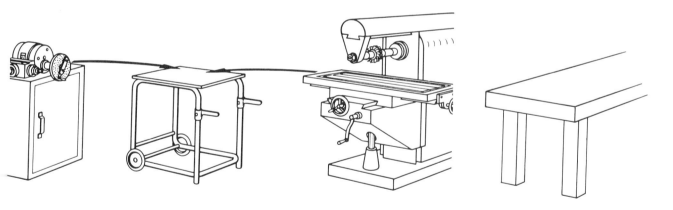

2.1.86 A hand controlled hoist or wheel stand in use

1.n Faults in marking out: their causes and correction

1.n.i Faults will show up in the product

fault	cause	correction
scribed lines not in the right places	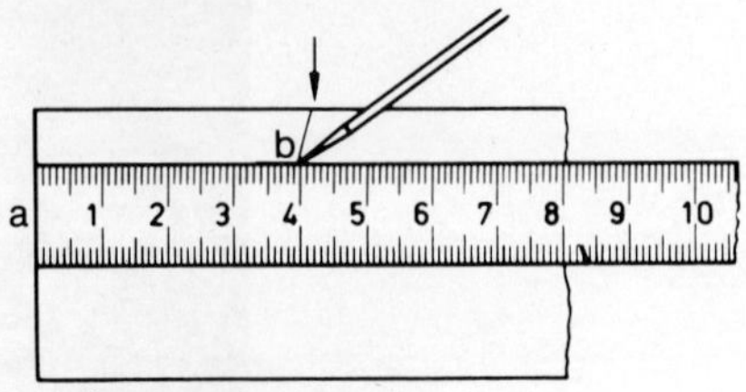 marking out tools used incorrectly in marking off the dimension	place the workpiece and the steel rule against a datum block 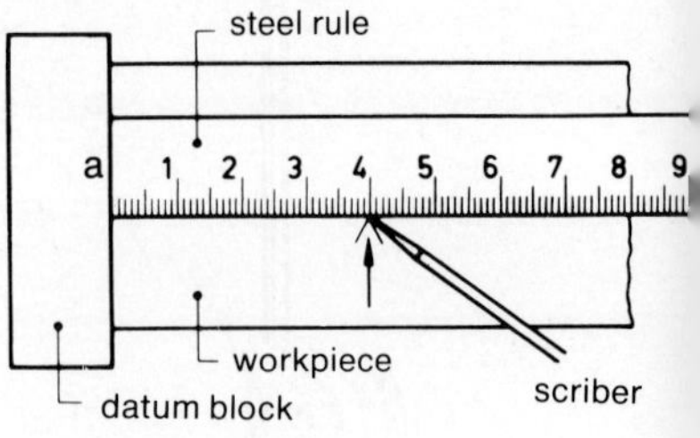

rule not square on at a; inaccurate measurement at b

steel rule — workpiece — datum block — scriber

| lines too thick | blunt scriber
scriber too soft
scriber too hard (point breaks) | sharpen it (grind)
harden it
temper it |

| straight lines not clear | scriber held incorrectly | hold scriber correctly |

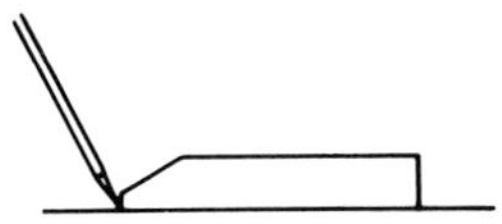

scriber protrudes too far and is not parallel to the surface plate

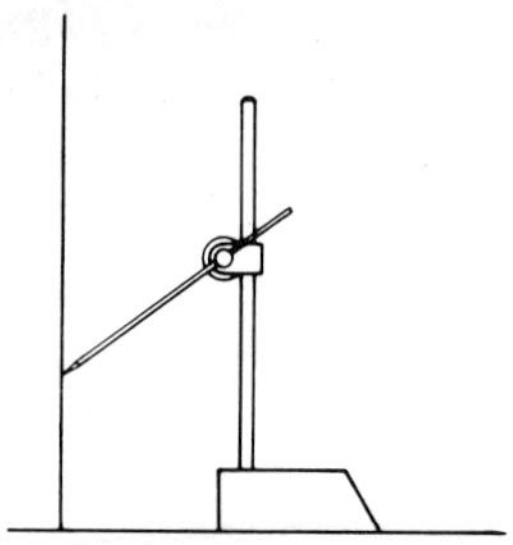

set scriber to protrude as little as possible and parallel to the surface plate

scriber bends

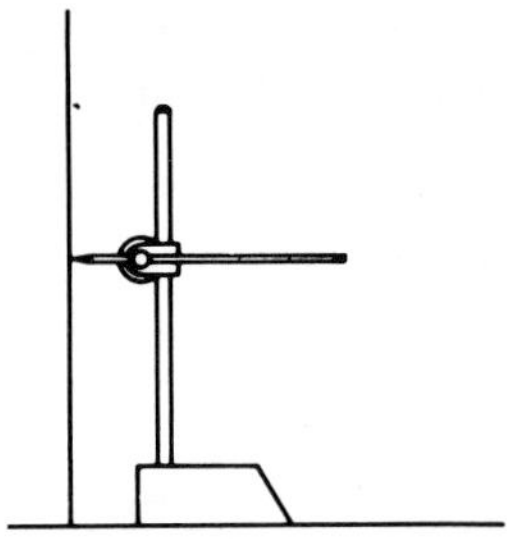

| centre marks beside the lines | centre punch placed wrongly to make the centre marks 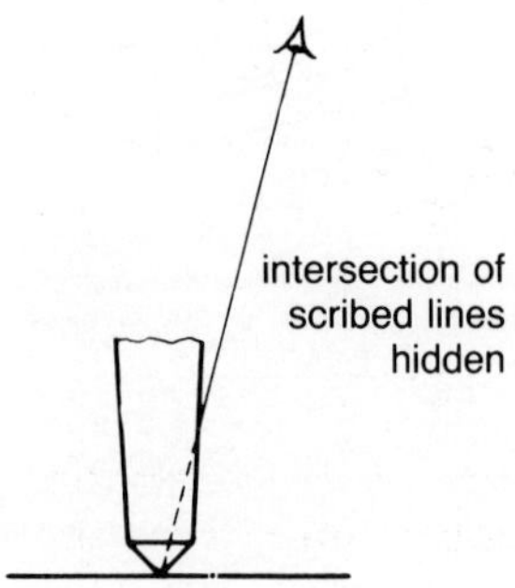 | support the hand holding the centre punch; hold the punch at an angle so that the point and the intersection of scribed lines remain visible: bring the punch upright and strike it. 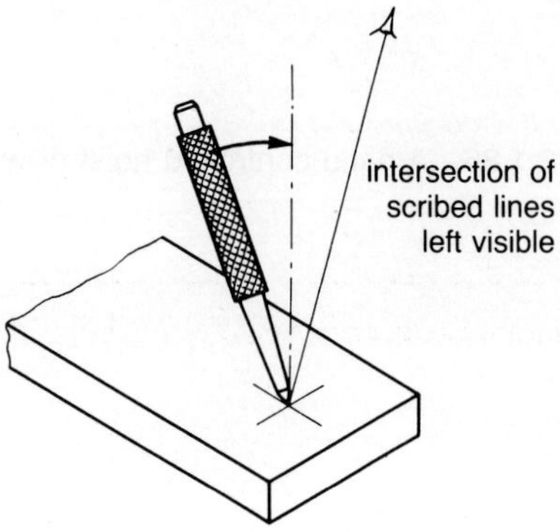 |

intersection of scribed lines hidden

intersection of scribed lines left visible

| protective coatings damaged | scriber far too hard | use soft scribers (pencils)on tinned and galvanised sheet as well as on aluminiur and aluminium alloy sheet |

1.n.i Faults will show up in the product

fault	cause	correction
circles and arcs not clear	dividers used incorrectly	press hard enough on the divider leg placed in the centre mark
	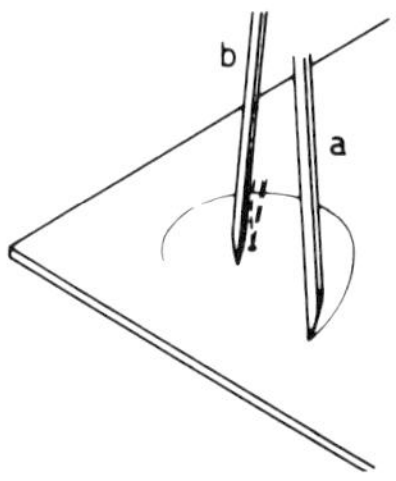 divider legs not of equal length	grind divider legs to equal length. points must lie flush (for drawing circles of less than about 12 mm diameter, however, divider legs are usually ground or adjusted to unequal lengths) 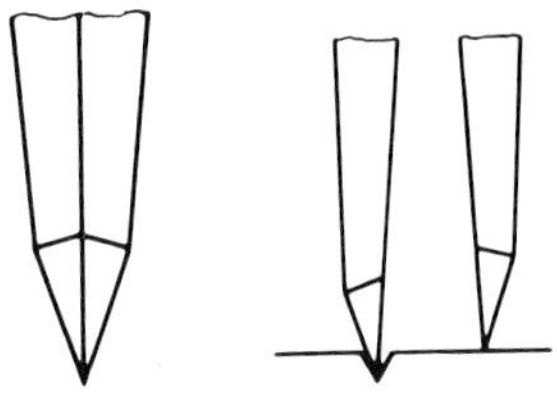

pressure on leg *a* is too great:
leg *b* wanders off the centre mark

| | dividers at wrong angle or centre mark wrong | marking out tools must be ground to the correct angle |

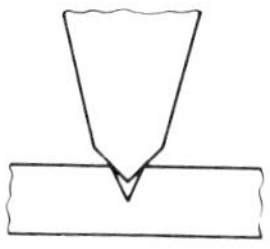

dividers point ground to wrong angle   centre mark wrong

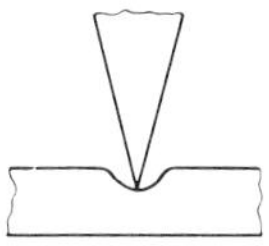

| material tears when folded | scriber too hard | use soft (brass) scriber |
| | scribed on the wrong side 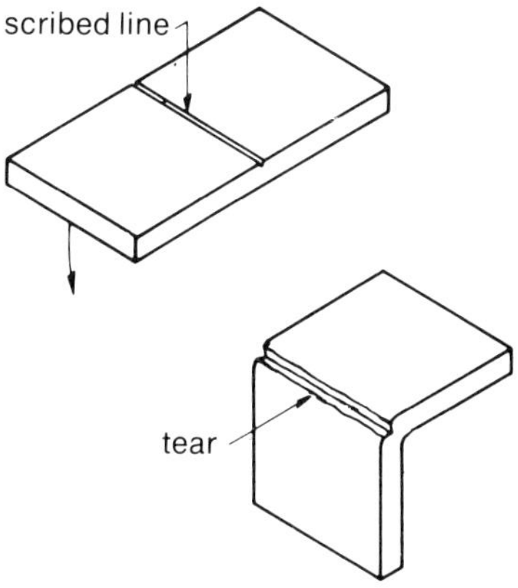 | when hard scribers are used, the scribed line should be on the inside of the fold 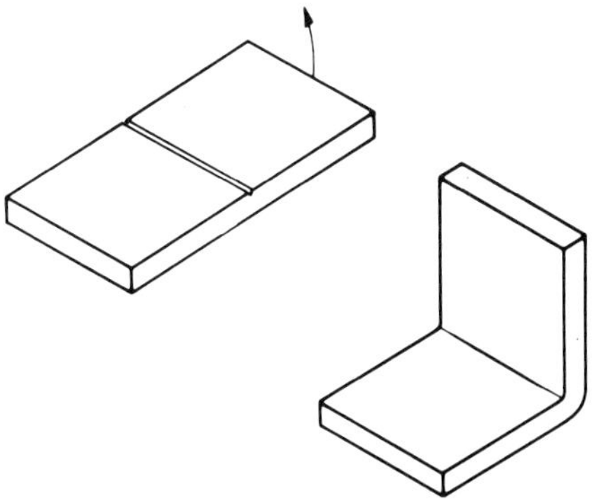 |

scribed line

tear

| inaccurate measurements | use of unsuitable rules | use only engineer's rules and steel metre tapes |

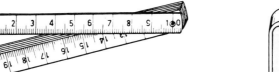

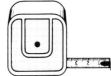

Acknowledgements

The publishers gladly record their thanks to Neill Tools Limited, Sheffield who kindly supplied photographs for inclusion in this book.

Technology of Skilled Processes

Basic Engineering
Competences 201

Measuring and
Marking Out

Practice and test questions

Published as a
co-operative venture
between
Stam Press Ltd

and

**City and Guilds of
London Institute**

Practice and test questions

The questions in this book are intended to help the student achieve and demonstrate a knowledge and understanding of the subject matter covered by this book. Accordingly, the questions follow the original chapter order, under the same headings. Finally there are questions spanning the chapters and approximating to the level of those in the relavent examination of the City and Guilds of London Institute.

FOR THE ATTENTION OF THE TEACHER AND THE STUDENT

The content of this book and the questions for the student have been carefully prepared by a group of special editors in co-operation with the City and Guilds of London Institute. We should like to draw your attention to the copyright clause shown at the beginning of the book, on this page and the following pages:

Australia	AEP, Blackburn (Melbourne)
België	Plantyn, Deurne (Antwerpen)
Belgique	Plantyn, Bruxelles
BRD	Stam, Köln
France	Casteilla/Educalivre, Paris
Great Britain	Hulton, Amersham
	Stam Press, Amersham
	Stanley Thornes, Cheltenham
Nederland	Educaboek, Culemborg
	Educa Int., Culemborg
	De Ruiter, Gorinchem
Suisse	Delta & Spes, Denges (Lausanne)

MEASURING   Name: _____ Class: _____ Number: _____

1 Introduction to Measurement

The following questions relate to the above subject, but what is said here about answering them *also* applies to similarly framed questions covering the later subject headings.

Questions, like *1–3*, require you to chose the correct combination of letters and numbers, as in the following example:

Example

Indicate which words (a–d) refer to 1 and which to 2:

a	short	1	length	1	**a, d**
b	square				
c	round	2	shape	2	**b, c**
d	long				

Other questions show a number of possible answers (mostly four, lettered a, b, c and d), but only *one* is correct. You are required to decide which it is and to circle the related letter. Thus, if you decide it is b you show it like this, ⓑ.

When you answer questions featuring drawings, their clarity can be increased considerably if you colour sections, angles and even important lines.

---

1 Indicate which words (a–d) refer to 1 and 2:

a straight   c curved   1 position   _____

b horizontal   d vertical   2 shape   _____

---

2 Indicate which words (a–e) apply best to 1, 2 and 3:

a cold

b quick   1 temperature   _____

c rough

d hot   2 speed   _____

e slow   3 finish   _____

---

3 Indicate which of the following statements apply to 1 and 2:

a an angle is precisely 90°

b an angle is slightly smaller than 90°

c an angle is slightly larger than 90°

1 the angle is not a right angle (not 'square')   _____

2 the angle is a right angle ('square')   _____

---

4 Show which of the indicating measuring instruments listed is non-adjustable:

a a vernier caliper

b a steel rule

c a micrometer

d a dial test indicator

---

5 Show which of the adjustable indicating instruments is suitable for measuring length:

a an outside caliper

b a straight edge

c a micrometer

d a bevel protractor

---

6 A screw pitch gauge as shown is used to compare:

a the thread pitch and form

b the thread angle

c the screw thickness

d the screw length

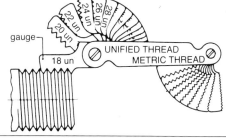

---

7 The micrometer shown can be used to check:

a flatness

b squareness

c run-out

d parallelism

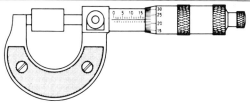

---

85   © Stam Press Ltd. Amersham, 1986

8 Deviation of an angle in degrees can be determined by:
- a   a bevel gauge
- b   a bevel protractor
- c   a mitre square
- d   a try square

9 In checking for angle deviations with a dial test indicator as shown, deviation may be expressed in:
- a   degrees
- b   degrees, minutes and seconds
- c   millimetres
- d   parts of a millimetre

cylindrical square   dial test indicator set to zero   surface plate   touching   workpiece   reading of deviation   touching

10 Radial run out can be measured with a:
- a   snap gauge
- b   dial test indicator
- c   vernier caliper
- d   micrometer

11 Show which type of indications (a, b, c, d or e) occur when using items 1, 2 or 3 to check flatness.
- a   shining spots
- b   light gap
- c   right-wrong
- d   parts of a millimetre
- e   the magnitude of deviation

1   surface plate   _____

2   straight edge   _____

3   dial test indicator   _____

12 In checking a right angle, which of the following are appropriate to 1, 2 and 3.
- a   parts of a millimetre
- b   degrees, minutes and seconds
- c   right-wrong
- d   not square

1   try square   _____

2   bevel protractor   _____

3   cylindrical square and
    dial test indicator   _____

13 The shaft shown is to be checked and/or measured on the items mentioned below the drawing. Name the most suitable measuring tool for 1, 2, 3 and 4.

1   diameter   2   length   3   run-out   4   parallelism

_____   _____   _____   _____

Give short and clear answers to the following questions.
Example
Question: State the SI unit of length.
Answer: The metre.

14 Name THREE non-indicating measuring tools.   _____
_____
_____

15 Name THREE measuring tools which can be used to determine the magnitude of a linear deviation from a stated dimension.   _____
_____
_____

16 Name FOUR indicating measuring tools.   _____
_____
_____

17 State if the metre is a base unit or a derived unit.   _____

18 State if the m^2 is a base unit or a derived unit.   _____

19 From which base unit is the unit of speed derived?   _____

20 a Give a multiple of a metre.   _____

    b Name TWO sub-multiple parts of a metre.   _____

2 Measurement of length

The following questions relate to the above subject. Except where otherwise indicated, the correct answers should be given as explained on page 85.

1 The most accurate of the listed instruments is:
  a a vernier caliper
  b a dial test indicator
  c a steel rule
  d a micrometer

2 The graduation of a steel rule is:
  a smaller than the measuring accuracy
  b usually   1 mm
  c usually 0.5 mm
  d larger as the rule becomes longer

3 From the figure allocate the correct letters to the numbers 1 to 8.
  a depth measuring rod
  b moving jaw
  c fixed jaw
  d clamp screw
  e graduated beam
  f slide assembly with vernier

1 _____   2 _____   3 _____

4 _____   5 _____   6 _____

7 _____   8 _____

4 Fill in the correct answers for the vernier scales in figures 1, 2 and 3.
  a 0.05 mm vernier scale
  b 0.02 mm vernier scale
  c 0.001 in vernier scale

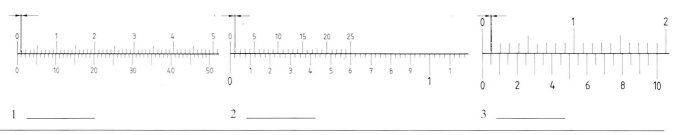

1 _____   2 _____   3 _____

5 A universal vernier caliper is used for:
  a checking squareness
  b checking flatness
  c measuring lengths having a tolerance of 0.25 mm or more
  d measuring lengths with a tolerance of 0.1 mm

© Stam Press Ltd. Amersham, 1986

6 In figures 1, 2 and 3 indicate which are the correct and incorrect methods of measuring the depth of a hole.
   a   right
   b   wrong

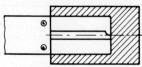

1 _____   2 _____   3 _____

7 The use of a vernier caliper is shown in the figures below. These figures are numbered 1, 2, 3 and 4, but not in the
   normal operational order. Indicate the correct order using the letters a, b, c and d.

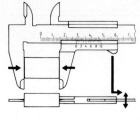   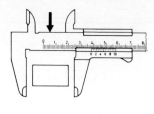   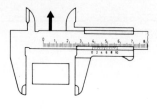   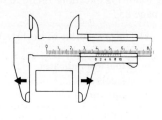

1 _____   2 _____   3 _____   4 _____

8 In measuring with a vernier caliper, the measuring force
   should be:
   a   adapted to the shape of the work
   b   greater as the measuring length becomes greater
   c   constant and light
   d   constant but heavy

9 Four dimensions are designated in four different drawings   a   pocket tape
   as shown. The scale in which these dimensions are drawn   b   steel rule
   is also different. Fill in the right measuring instrument to   c   measuring tape
   measure each of the dimensions.   d   vernier caliper

   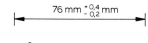   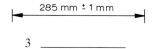   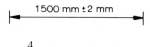

1 _____   2 _____   3 _____   4 _____

The questions *10–14* refer to the micrometer shown.
Its parts are numbered and their names are:
a   frame
b   thimble
c   barrel
d   spindle bearing
e   spindle thread
f   stem
g   ratchet
h   anvil
i   thimble adjusting nut
j   locking lever
k   spindle
l   screw adjusting nut

In questions *10–14* enter these names, relating them correctly to the figure shown.

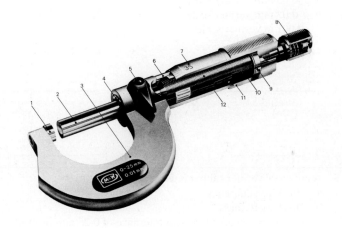

10 1 = _____   2 = _____

11 3 = _____   4 = _____

12 5 = _____   6 = _____

13 7 = _____   8 = _____   9 = _____

14 10 = _____   11 = _____   12 = _____

15 The range of an external micrometer is:
   a   usually 25 mm
   b   0–50 mm
   c   0–100 mm
   d   0–200 mm

16 The thimble of a micrometer is usually graduated into:
   a   25 divisions
   b   50 divisions
   c   100 divisions
   d   200 divisions

17 If a micrometer is provided with a vernier scale, readings
can be taken to an accuracy of:
   a   0.1 mm
   b   0.05 mm
   c   0.001 mm
   d   0.0001 mm

18 For each of the measuring instruments 1, 2, 3 and 4 indicate whether a, b, c or d is correct:
   a   scale value 1   mm
   b   scale value 0.05 mm
   c   scale value 0.02 mm
   d   scale value 0.001 mm

   1   vernier caliper with a 1/50 vernier _____
   2   vernier caliper with a 1/20 vernier _____
   3   steel rule   _____
   4   micrometer with a vernier   _____

19 Give the correct answer for figures 1, 2, 3 and 4.
The diameter D shown in mm in figures 1–4 can be measured
with a:
   a   micrometer with a vernier scale
   b   micrometer
   c   vernier caliper
   d   steel rule

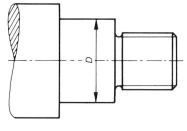
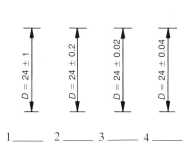

1 _____ 2 _____ 3 _____ 4 _____

20 Fill in the correct reading for figures 2–5 (see figure 1 for example).
   a   6.15 mm   c   6.65 mm   e   10.04 mm
   b   6.5 mm   d   14.64 mm

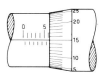

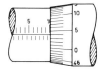
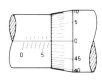

   1 = a = 6.15 mm   2 = _____   3 = _____   4 = _____   5 = _____

21 A micrometer usually has:
    a   a thread pitch of 0.5 mm and a range of 50 mm
    b   a thread pitch of 0.5 mm and a range of 25 mm
    c   a thread pitch of   1 mm and a range of 25 mm
    d   a thread pitch of   1 mm and a range of 50 mm

22 In two-point measurement the smallest inside dimension
which can be measured with a micrometer is:
    a   6 mm
    b 25 mm
    c   8 mm
    d 30 mm

23 In three-point measurement the smallest inside dimension which can be measured with a micrometer is:
    a   6 mm
    b 25 mm
    c   8 mm
    d 30 mm

24 An inside micrometer for two-point measuring may be
tested for accuracy using:
    a   a dial indicating micrometer
    b   0.01 mm screw
    c   a checking ring
    d   a vernier micrometer

Questions 25–29 refer to dial test indicators; continue answering as explained on page 85.

25 For dial test indicators which of the following is correct?
    a   the smaller the range, the greater the accurary
    b   the greater the range, the greater the accuracy
    c   the smaller the range, the less accurate
    d   the accuracy is constant for any range

26 The stylus of a dial test indicator travels a certain distance
when a measurement is taken. Which of the following is
correct?
    a   the shorter the distance, the more accurate the measurement
    b   the greater the range, the greater the accuracy
    c   the smaller the range, the less accurate
    d   the accuracy is constant for any range

27 Select suitable graduation values (a, b or c) for the dial
test indicator ranges shown:
    a   graduation value 0.001   mm
    b   graduation value 0.01   mm
    c   graduation value 0.0005 mm

Range: 1 1–10 mm   _____
              2 up to 1 mm   _____
              3 up to 0.025 mm   _____

28 Fill in whichever is applicable:

    a   oversize
    b   undersize

1 _____   2 _____   3 _____   4 _____

29 Study figures 1–4 of question *28* and fill in whichever is applicable:

a shaft smaller

b shaft larger   Figure 1 _____

c hole larger   Figure 2 _____

d hole smaller   Figure 3 _____

e may be remachined to size

f a possibility of rejection   Figure 4 _____

Give short and clear answers to questions *30–76*.

Example

Question: State the measuring range of a steel rule commonly used.

Answer: Usually 300 mm.

---

30 State the usual graduation value of a steel rule. _____

31 State TWO factors to be observed when measuring with a steel rule. _____

32 State the advantage os using a datum when measuring with a steel rule. _____

33 State the expected accuracy of measurement when using a steel rule. _____

34 State one factor which governs the accuracy of the reading obtained from a vernier caliper. _____

35 State the readings from the vernier scales shown below.

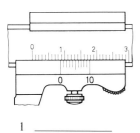 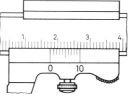 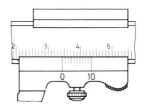 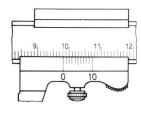

1 _____   2 _____   3 _____   4 _____

---

36 The figure shown is an enlarged vernier scale.
State how this has been divided.

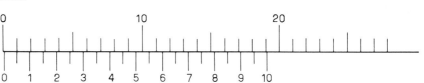

_____

---

37 State the readings from the vernier scales shown below.

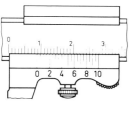 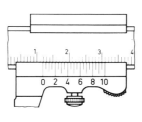 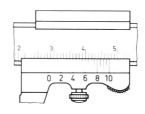 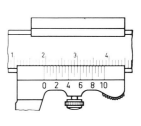

1 _____   2 _____   3 _____   4 _____

---

38 State two factors to be observed when measuring with a vernier caliper. _____

39 State the expected accuracy of measurement when a vernier caliper is used. _____

40 State the reading value of a micrometer. _____

41 State the range of a micrometer 25–50 mm. _____

42 State the range of a micrometer 50–75 mm. _____

43 State the usual range of a micrometer without accessories. _____

44 State why the range of a micrometer is limited. _____

45 State the usual thread pitch of the spindle of a micrometer. _____

46 State the number of divisions of the thimble of a micrometer. _____

47 State THREE types of micrometer. _____

48 State the function of a micrometer. _____

49 If a micrometer is provided with an additional 10-division vernier scale, state the degree of precision. _____

50 State the number of points of contact made by TWO different types of inside micrometer. _____

51 State the smallest size which can be measured with a two-point inside micrometer. _____

52 State the range of the smallest available two-point micrometer. _____

53 State the largest range of a two-point inside micrometer. _____

54 State the value of each division on the thimble of a three-point inside micrometer. _____

55 State the accuracy of measurement obtainable when using a two-point inside micrometer. _____

56 State the accuracy of measurement obtainable when using a three-point inside micrometer. _____

57 Name the equipment used to check the accuracy of the readings of an external micrometer. _____

58 Name the equipment used to check the accuracy of the readings of an inside micrometer. _____

Figures 2–4 relate to the micrometer in fig. 1 and give different readings. Questions *59–62* relate to these illustrations.

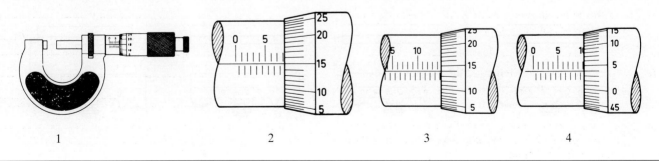

1   2   3   4

59 Name the kind of micrometer shown in fig. 1.   _____

60 State the reading shown in fig. 2.   _____

61 State the reading shown in fig. 3.   _____

62 State the reading shown in fig. 4.   _____

Figures 2–4 relate to the micrometer in fig. 1 and give different readings. Questions *63–65* relate to these illustrations.

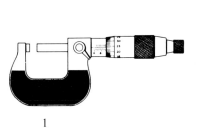
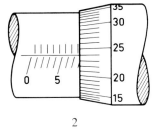
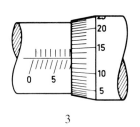
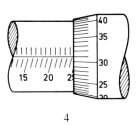

1   2   3   4

63 State the reading of fig. 2.   _____

64 State the reading of fig. 3.   _____

65 State the reading of fig. 4.   _____

66 Write the reading under the figures 1–7.

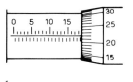
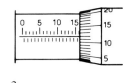
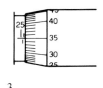

1 _____   2 _____   3 _____

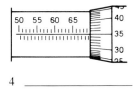
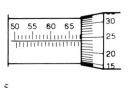
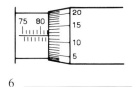
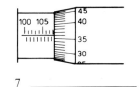

4 _____   5 _____   6 _____   7 _____

67 State the usual measuring ranges of four different dial test indicators.   _____
_____

68 State the value of one division on the scale of an indicator with a range of 0–10 mm.   _____
_____

69 State the connection between the measuring range and the accuracy of dial test indicators.   _____
_____

70 What connection is there between the distance travelled by the stylus and the accuracy of measurement?   _____
_____

71 State the accepted tolerance of measurements obtainable with a dial test indicator having a dial graduation of 0.01 mm and a range of 10 mm.   _____

72 State the accepted tolerance of measurements obtainable with a dial test indicator having a dial graduation of 0.001 mm and a range of 1 mm.   _____

73 State the accepted tolerance of measurements obtainable with a dial test indicator having a dial graduation of 0.0005 mm and a range of 0.025 mm.

_____

74 Name the additional accessory required to measure a dimension with a stand-mounted dial test indicator.

_____
_____

75 State THREE uses for a dial test indicator.

_____
_____

76 Which measuring instrument is used to measure:
a   an external length of 5060 mm ± 2 mm?
b   an external length of 1750 mm ± 2 mm?
c   a shaft diameter of 48 mm ± 0.4 mm?
d   a shaft diameter of 52 mm ± 0.03 mm?
e   a hole dimension of 20 mm ± 0.1 mm?
f   a hole dimension of 24 mm ± 0.02 mm?
g   a run out between 0 and 0.24 mm?
h   a depth of 6 mm $\pm \begin{smallmatrix} 0.02 \\ 0 \end{smallmatrix}$ mm?

_____

Questions *77–81* require completion of the figures:
Example
The figure a illustrates a depth to be measured with a vernier caliper: it should be completed as shown in figure b.

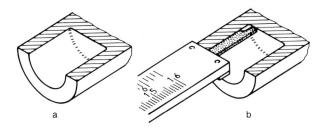

a   b

77 From the list below mark the different parts in the figure by inserting the appropriate numerals in the circles.
　1   anvil
　2   spindle
　3   ratchet
　4   spindle bearing
　5   thimble
　6   stem
　7   spindle thread
　8   frame
　9   thimble adjusting nut
　10   barrel
　11   locking lever
　12   screw adjusting nut to remove play

78 Figure 1 represents the checking of the vertical positions of a plane. The dial test indicator is held in the spindle of a drilling machine. Complete the drawing.

79 In figure 2 the position of a plane V_1 is to be checked with a dial test indicator held in the spindle of a milling head.
Complete the drawing.

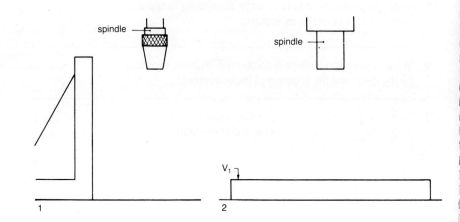

spindle

spindle

V_1

1   2

94

80 In figure 3 a machine vice is to be aligned using a dial test indicator. The indicator is held in the spindle of a milling machine. Complete the drawing.

81 In figure 4 the horizontal position of a shaft is to be checked. The dial test indicator is mounted on a stand. Complete the drawing.

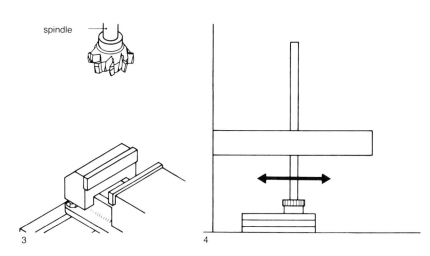

3 Angle measuring equipment

The following questions relate to the above subject. The correct answers to questions *1–11* should be indicated as explained on page 85.

1 A try square with a blade length of 75 mm and a stock
length of 50 mm is designated:
a   74 + 50
b   75 − 50
c   75 × 50
d   50 × 75

2 A bevel edge square:
a   indicates the deviation in units when the angle is smaller than 90°
b   indicates the deviation in units when the angle is larger than 90°
c   indicates the deviation in units when the angle is smaller or larger than 90°
d   does not indicate the deviation in units

3 Inspection squares are used for:
a   testing workshop squares
b   laying out right angles
c   checking ground workpieces
d   checking milled workpieces

4 In testing a square as in figure 1 and figure 2, a deviation is
expressed in:
a   hundredths of millimetres
b   millimetres
c   degrees
d   seconds

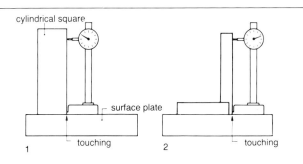

5 The unit of angle is:
a   90°
b   1 minute
c   1 degree
d   1 second

6 Complementary angles add up to:
   a   360°
   b   270°
   c   180°
   d   90°

7 Supplementary angles add up to:
   a   360°
   b   270°
   c   180°
   d   90°

8 A bevel protractor without vernier measures to an accuracy of:
   a   1°
   b   0.5°
   c   5′
   d   2′30″

9 A bevel protractor with a 12 division vernier can be read to an accuracy of:
   a   1°
   b   0.5°
   c   5′
   d   2′30

10 A bevel protractor with a 24 division vernier can be read to an accuracy of:
   a   1°
   b   0.5
   c   5′
   d   2′30″

Give short and clear answers to questions *11–24*.

11 State what cylindrical squares are used for. _____
_____

12 How many degrees are there in a right angle? _____

13 How many seconds are there in a degree? _____

14 State the sum of two angles which are the complement of each other. _____

15 State the sum of two angles which are the supplement of each other. _____

16 State how many degrees are divided by a bevel protractor vernier scale which is graduated in 12 divisions. _____

17 State the part of a degree represented by one line of a 12 division vernier scale. _____

18 With a bevel protractor having a 12 division vernier scale state the difference between 1 division on the scale and 2° on the protractor dial. _____

19 State the accuracy to which a reading can be taken on a protractor with a 12 division vernier. _____

20 State how many degrees are divided by a bevel protractor vernier scale graduated in 24 spaces. _____

21 State the value of one division on a 24 division vernier scale. _____

22 With a 24 division vernier scale state the difference between 1 division on the scale and 1° on the protractor dial.

_____

23 State the accuracy to which a reading can be taken on a protractor with a 24 division vernier.

_____

24 State whether the vernier of a protractor runs in only one direction from zero, or both ways.

_____

In questions 5–29 the space below or beside the question should be used for the answer and to complete the figure

Example
Question: state the reading shown in the figure.
Answer: 31°25'

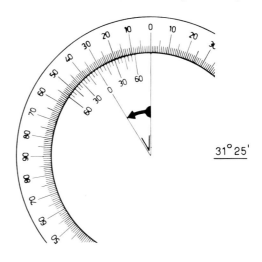

31°25'

25 Indicate in figure 1:
  a a right angle with a red arrow
  b an angle of 180° with a blue arrow

26 Draw an angle of 30° in figure 2 and mark that angle and its complement in different colours.

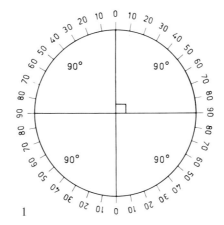

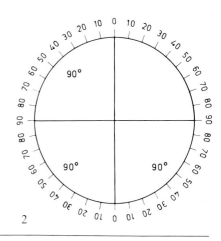

27 Draw an angle of 70° in figure 1 and mark that angle and its supplement in different colours.

28 Draw an angle of 130° in figure 2 and mark that angle and its supplement in different colours.

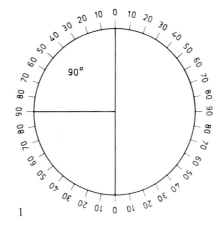

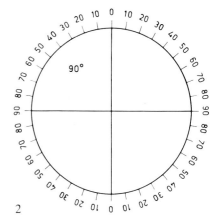

29 Enter the names of the numbered parts in the figure
shown

1 = _____

2 = _____

3 = _____

4 = _____

5 = _____

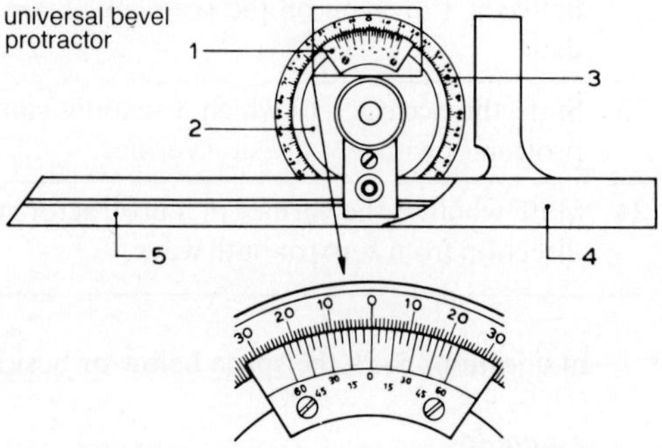

universal bevel
protractor

Questions *30–34* following relate to the reading of a universal bevel protractor: they should be answered as indicated individually.

---

30 The figure in this question has been prepared as an
example.
a   The angle α to be measured is hatched.
b   The angle α to be read is hatched similarly.
c   Complete the sentences.

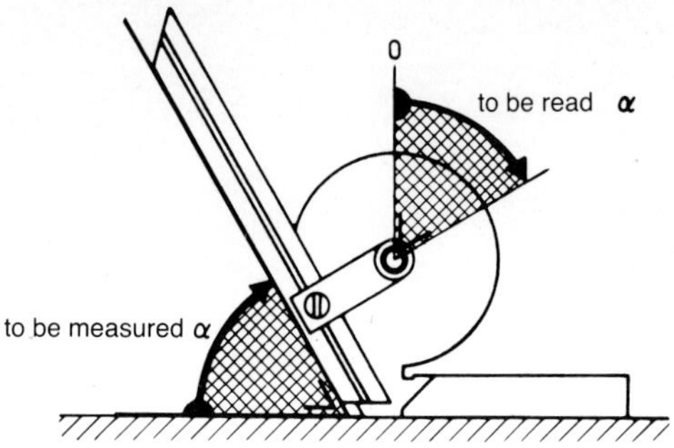

The required angle is_____

Fractional parts of a degree are read from:

_____

to be read **α**

to be measured α

---

31 a   Colour the angle β to be measured.
b   Mark by directional arrows the two angles which
when subtracted from each other will equal the angle
β to be read.
c   Mark the angle β to be read in the same colour as the
angle β to be measured.
d   Complete the sentences.

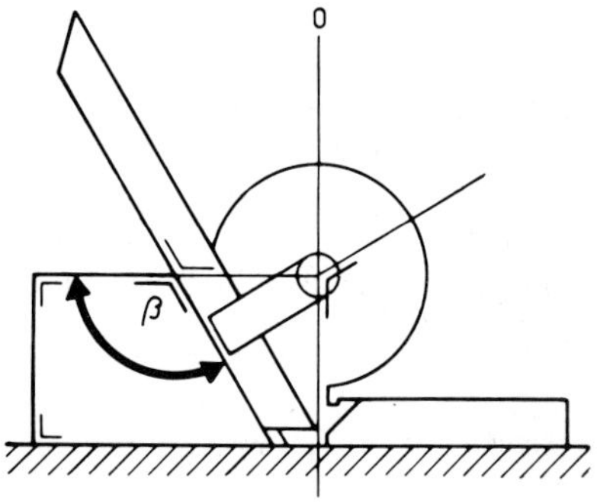

The required angle is: _____

Fractional parts of a degree

are read from: _____

---

32 a   Indicate the angle γ to be measured by an arrow
headed arc.
b   Colour this angle.
c   Mark by directional arrows the two angles which when
added together will equal the angle γ to be read.
d   Mark the γ to be read in the same colour as the angle
to be measured.
e   Complete the sentences.

The required angle is: _____

Fractional parts of a degree

are read from: _____

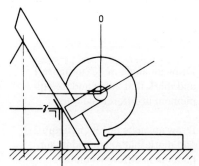

---

33 a Indicate the angle δ to be measured by an arrow
   b Colour the angle δ to be measured.
   c Mark by directional arrows the two angles which when
      subtracted from each other will equal the angle δ to be
      measured.
   d Mark the angle δ to be read in the same colour as the
      angle δ to be measured.
   e Complete the sentences.

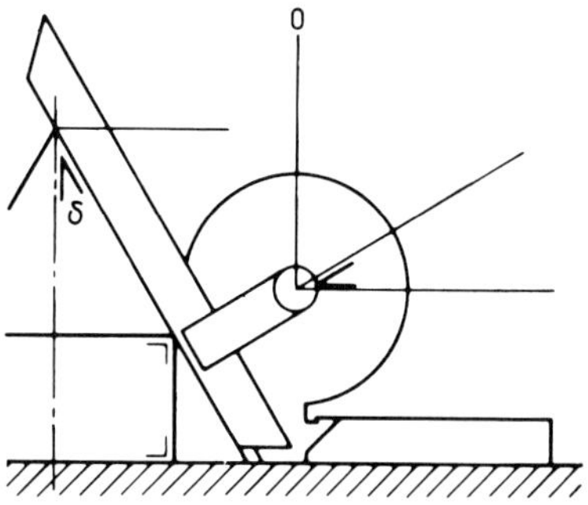

The required angle is: _____

Fractional parts of a degree

are read from: _____

34 The indicated angle in questions *27–33*
   is always 58°30′.
   How many degrees are α, β, γ and δ?

α = _____

β = _____

γ = _____

δ = _____

4 Accuracy of measurement

The following questions relate to the above subject. Except where otherwise required, the correct answers to questions should be indicated as explained on page 85.

1 It is correct to say that:
   a the highest possible precision will be cheapest in the long run.
   b the highest possible precision is desirable in any circumstances.
   c a higher accuracy than necessary for proper functioning increases production costs.
   d a higher accuracy than necessary for proper functioning will not increase production costs.

2 Indicate which applies to figure 1 and figure 2.
   a Symmetrical dimensional deviation
   b Asymmetrical dimensional deviation

$45_{-0.4}^{0}$

45 ± 0.2

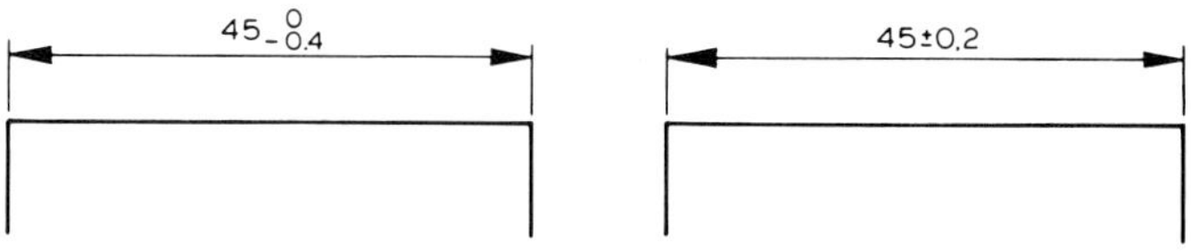

1 _____

2 _____

3 In a symmetrical dimensional deviation:

   a the positive deviation is equal to the negative deviation.

   b the positive deviation is larger than the negative deviation.

   c the positive deviation is smaller than the negative deviation.

   d the nominal size is larger than the ideal size.

4 Relatively, the most accurate
   dimensional deviation represented in figure a, b, c or d?
   The nominal sizes are expressed
   in millimetres.

a 5000 ± 1

b 30 ± 0.01

c 400 ± 0.1

d 2 ± 0.001

© Stam Press Ltd. Amersham, 1986

5   Four measuring instruments are listed.
Indicate the correct tolerance for each instrument:

1   steel rule   _____

2   micrometer   _____

3   dial test indicator _____

4   vernier caliper   _____

6   State TWO reasons for allowing dimensional deviations.

_____

_____

_____

_____

7   Enter on the dimension line a dimension of which:
a   the nominal size is 70 mm
b   the positive and the negative dimensional deviation
are both 0.1 mm

8   Enter on the dimension line a dimension of which:
a   the nominal size is 70 mm
b   the positive deviation is + 0.07 mm
c   the negative deviation is − 0.03 mm

9   Is there a difference between nominal size and the mean
size in the designations a and b?
If in either case the answer is yes, state the difference.

a   60 mm ± 0.05 mm? _____

b   50 mm ± $\frac{0.15}{0.05}$ mm? _____

10   State THREE factors which influence the accuracy of
precise measurement.

_____

_____

_____

_____

Questions *11–20* deal with terms like mean size, positive and negative deviations, limits and tolerance: give short and clear answers.

11   The high and low limits of a workpiece are 44.8 mm and
45.0 mm. What is the mean size?   _____

12   State the mean size of the dimension 25 mm $^{+\ 0.2}_{\ \ \ 0}$ mm.   _____

13   State the mean size of the dimension 25 mm $^{\ \ \ 0}_{-\ 0.2}$ mm.   _____

14   State the positive deviation if the diameter of a shaft is
given as 48 mm $^{\ \ \ 0}_{-\ 0.02}$ mm.   _____

15   State the negative deviation if the diameter of a shaft is
given as 48 mm $^{+\ 0.02}_{\ \ \ 0}$ mm   _____

16   With reference to the dimension shown, state: 60 $^{+0.3}_{-0.1}$
a   the tolerance
b   the limits of size   _____

17   A dimension is designated 15 mm $^{+\ 0.03}_{-\ 0.01}$ mm.
State:
a   the limits of size   _____

b   the tolerance   _____

18 With reference to the dimension shown, state:
    a the mean size
    b the tolerance
    c the limits of size

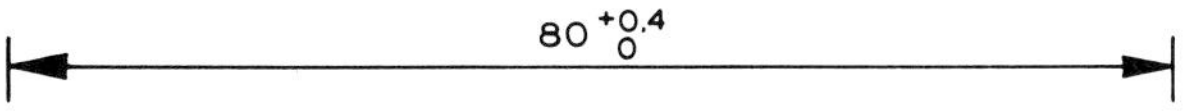

$$80^{+0.4}_{\ 0}$$

---

19 A dimension is designated 50 mm $^{\ \ 0}_{-0.04}$ mm.
State:
    a the positive deviation

    b the negative deviation

    c the limits of size

    d the mean size

---

20 With reference to the dimension shown, state:

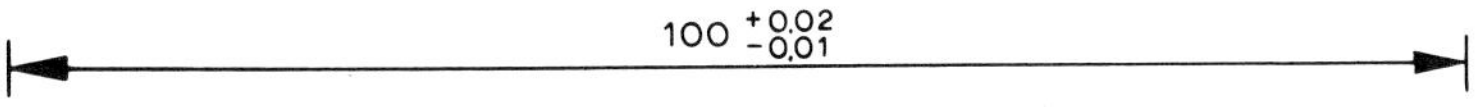

$$100^{+0.02}_{-0.01}$$

    a the positive deviation

    b the negative deviation

    c the tolerance

    d the limits of size

The correct answers to questions *21–35* should be indicated as explained on page 85.

---

21 For uniform accuracy measurements should be taken
when both workpiece and measuring tool are at:
    a 0°C
    b the same temperature
    c 20°C

---

22 The possibility of making a reading error is smallest in figure 1, A, B or C.

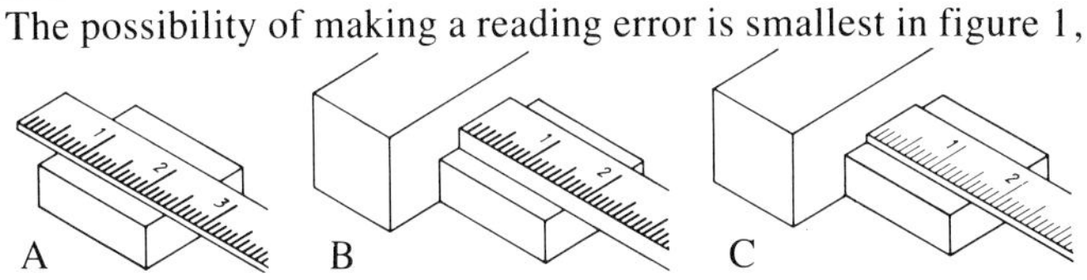

A   B   C

---

23 The ratchet of a micrometer serves to ensure:
    a a constant measuring pressure
    b a constant measuring force
    c that the measuring pressure is reduced for smaller dimensions
    d that the measuring force is reduced for smaller dimensions

---

24 Referring to each figure shown,
indicate whether the dimension
obtained will be:
    a right
    b too small
    c too large

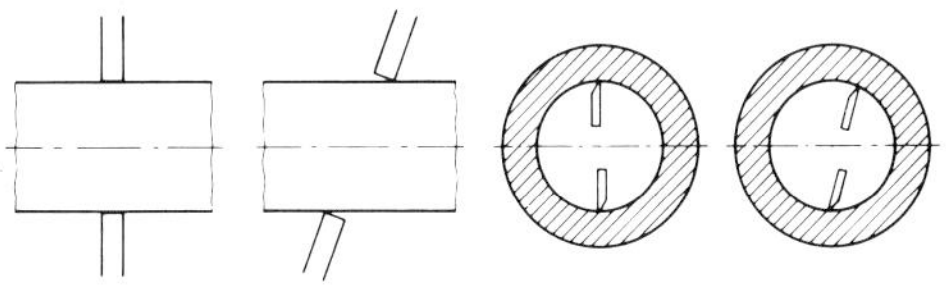

1 ____   2 ____   3 ____   4 ____

---

25 A specified dimension is:
    a a dimension shown on the drawing
    b always a length
    c always an angle
    d always an inside dimension

---

26 A measured reading is:
    a a dimension on a drawing
    b a dimension indicated by a measuring instrument
    c a true dimension which is read from an instrument
    d a dimension read from a measuring instrument

---

27 In measuring, indicated size is:
    a the data on the drawing
    b the length or angle indicated by the measuring equipment
    c the dimension which is read from the measuring equipment
    d the limit of size

28 The mean size is:
   a larger than the nominal size
   b smaller than the nominal size
   c equal to the nominal size
   d halfway between the limits of size

29 The measured diameter of a shaft at 37°C is 76.02 mm
   (micrometer 20°C). It is cooled to 20°C (micrometer temperature unchanged) and measured again. The measured
   diameter of the shaft is then:
   a unchanged
   b less than 76.02 mm
   c more than 76.02 mm

30 A micrometer is used to measure a bronze shaft. Both
   have a temperature of 37°C. The measurement taken is
   80.09 mm. Bronze has a larger coefficient of expansion
   than steel. The shaft is measured again after its temperature and that of the micrometer has cooled down to
   20°C. The measurement will then be:
   a 80.09 mm
   b larger than 80.09 mm
   c smaller than 80.09 mm
   d as large as that at 37°C

31 A diameter on a drawing is 36 mm ± 0,01 mm. The most suitable tool to measure that dimension is:
   a a dial test indicator
   b a micrometer
   c a steel rule
   d a vernier caliper

32 In order to ensure that the true dimension lies within the
   limits of size a margin of accuracy is introduced. This is
   based on the workpiece tolerance specified and is a number by which the tolerance must be multiplied.
   It is usually:
   a 0.25–0.50
   b 0.10–0.25
   c 0.50–0.10
   d 3.00–5.00

33 The length of a piece of material is designated 2000 mm ± 1 mm.
   The only unacceptable length listed is:
   a 1995.5 mm
   b 2000   mm
   c 2000.5 mm
   d 2001   mm

34 A dimension on a drawing is 40 mm $^{+\,0.05}_{\quad 0}$ mm.
   Using an accuracy factor of 0.1, the accuracy of the measuring instrument to be used should be:
   a 0.25 mm
   b 0.125 mm
   c 0.005 mm
   d 0.01 mm

35 A dimension of 30 mm $^{+\,0.2}_{-\,0.1}$ mm has to be measured.
   Using an accuracy factor of 0.1 the accuracy of the measuring instrument to be used should be:
   a 0.6 mm
   b 0.1 mm
   c 0.3 mm
   d 0.03 mm

MEASURING   Name: _____ Class: _____ Number: _____

36   Name FOUR factors which affect the accuracy of measurement

_____

_____

37   State the internationally agreed temperature at which precise measurements should be taken.

_____

_____

38   State the function of the ratchet of a micrometer.

_____

39   State TWO causes of reading errors.

_____

_____

40   State the maximum and minimum dimensions in the figure below.

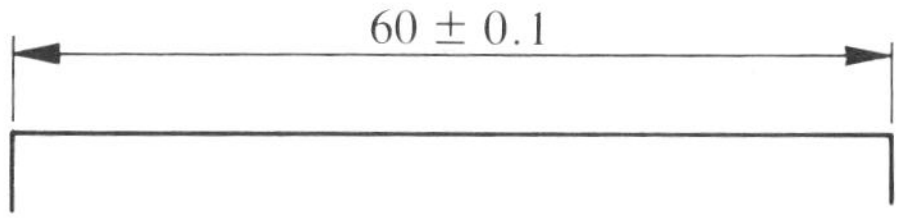

60 ± 0.1

_____

_____

41   Explain what is meant by the term 'the accuracy of a measurement'.

_____

_____

42   State the meaning of the term 'graduation value' of a measuring instrument.

_____

_____

43   State the meaning of the term 'the range' of an instrument for length measurements.

_____

_____

_____

44   State the normally accepted ratio between the instrument accuracy and workpiece tolerance.

_____

45   State TWO rules to be observed when taking accurate measurements.

_____

_____

46   A dimension of 88 mm $\pm$ 0.2 mm has to be measured. The accuracy of measurement must be 0.2 of the specified tolerance. The accuracy obtainable using a vernier caliper is $\pm$ 0.05. Would a measurement taken with this vernier caliper meet the requirements? Give a reason for your answer.

_____

_____

_____

47   Would a micrometer with a range of 75–100 mm and the accuracy of $\pm$ 0.01 mm meet the requirements of question *46*?

_____

_____

5 Determination of position, level and size

The following questions relate to the above subject. The correct answers to questions *1–6* should be indicated as explained on page 85.

1 The most widely used surface plates are made of:
   a   cast steel
   b   cast iron
   c   low carbon steel
   d   high carbon steel

2 The principal dimensions of a surface plate are:
   a   the length and the thickness
   b   the length only

3 A square block level has usually:
   a   one measuring surface
   b   three measuring surfaces
   c   two measuring surfaces
   d   four measuring surfaces

4 The sensitivity of a spirit level is expressed in:
   a   millimetre per metre for each vial division
   b   millimetre per metre
   c   millimetre for each vial division
   d   metres for each vial devision

5 The principal dimensions of a feeler gauge are:
   a   the width and thickness
   b   the width and length
   c   the thickness and length
   d   the length

6 An example of an indicating instrument is:
   a   a square
   b   a cylindrical square
   c   a spirit level
   d   a bevelled straight edge

7 An example of a non-indicating instrument is:
   a   a square
   b   a spirit level
   c   an optical protractor
   d   a vernier protractor

Give short and clear answers to the following questions.

8 State TWO materials from which surface plates may be made.

9 State which are the principal dimensions of a surface plate.

10 State how the accuracy of a spirit level is expressed.

11 State TWO uses of square block levels.

12 Explain why, as shown in this end elevation, spirit levels may have a vee in the base.

13 State TWO uses of feeler gauges

MARKING OUT   Name: _____ Class: _____ Number: _____

The following questions relate to the above subject. Questions *1–20* refer to marking out and marking tools and except where otherwise required, the correct answers should be indicated as explained on page 85.

Example
Question: Indicate which one of the following requires more marking out than the others

Answer: a   turning
          b   milling
          c   drilling
          d   surface grinding

---

1 A steel scriber is used to draw a marking out line. This line is:
   a   thick and easy to remove
   b   thin and sharply defined
   c   thick and difficult to remove
   d   thick and sharply defined

---

2 A soft pencil is used to draw marking lines on aluminium sheet. These lines are:
   a   thin and sharply defined
   b   much thinner than a scribed line and lie on the surface
   c   thicker than a scribed line and usually easy to remove
   d   thick and difficult to remove

---

3 For 1 and 2 indicate which of the items a–f are appropriate.
   a   surface plate
   b   scriber
   c   trammels
   d   steel rule
   e   angle plate
   f   pocket steel tape

   1   A support or datum face is
       provided by: _____

   2   Marking lines can be drawn with:
       _____

---

4 Indicate which of the following is used to draw lines parallel to a surface plate:
   a   an angle plate
   b   a surface gauge
   c   a box square
   d   a slip gauge

---

5 Parallel lines along the length of a shaft are drawn with the aid of:
   a   a scribing gauge
   b   dividers
   c   odd-leg calipers
   d   a box square

---

6 Parallel lines on angled sections may best be drawn with the aid of:
   a   a scribing gauge
   b   dividers
   c   odd-leg calipers
   d   a box square

---

7 Indicate which one of the listed items are used to locate the centre of the end of a round shaft.
   a   a surface gauge
   b   odd-leg calipers
   c   a centre square
   d   a box square
   e   a Vee-block and a surface gauge

© Stam Press Ltd. Amersham, 1986

8

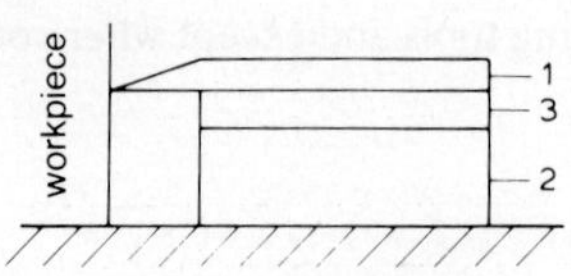

a 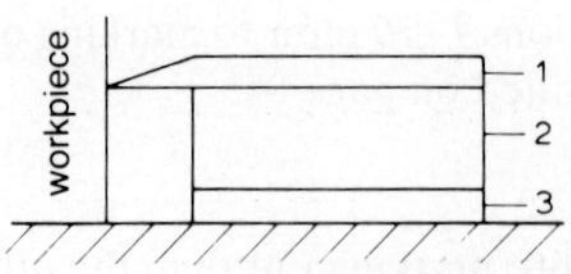   b 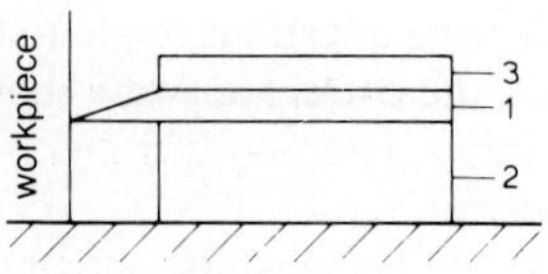   c

The above figures show how lines may be drawn with the aid of slip gauges:
1 = scribing block, 2 = slip gauge, 3 = protection slip
Colour the protection slip and circle the letter of the figure in which the gauges are stacked in the correct order.

9   The lines 1 and 2 on the figure, perpendicular to the sur-
       face plate, are drawn with a scriber along the edge of a:
       a   flat try-square
       b   flat try-square with stock support
       c   square with stock and blade
       d   mitre square

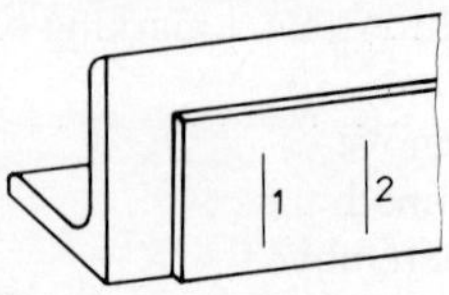

10 The lines 1 and 2 on the figure perpendicular to the side of
       the workpiece can be drawn along the edge of a:
       a   flat try-square
       b   flat try-square with stock support
       c   square with stock and blade
       d   mitre square

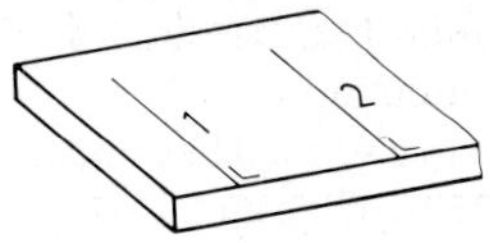

Give short and clear answers to questions *1–20*

Example
Question: State what material angle plates and Vee-blocks are made of.
Answer: Vee-blocks are made of cast iron.

11 State TWO reasons for marking
       out a component.

12 State why a marking medium
       may be used on workpieces to be
       marked out.

13 State a common use for a surface
       gauge.

14 State ONE advantage of mark-
       ing out equipment which can be
       read directly.

15 State ONE advantage a granite
       surface plate has over a cast iron
       surface plate.

16 State ONE advantage a carbide
       tipped scriber has over one
       which is entirely made of tool
       steel.

17 State TWO examples of the use
       of a spirit level. Illustrate your
       answer with simple sketches.

18 State why a steel scriber should
       not be used to mark a bending
       line on steel plate.

19 Explain why lines scribed on aluminium plate may have harmful consequences.

_____

_____

_____

20 Complete the table below by writing at least TWO answers in each column.

support or datum faces are/may be provided by:	Typical marking out equipment for straight or parallel lines is:	Typical marking out equipment for circles or parts of circles is:	Typical equipment for marking out angles is:

Questions *21–25* refer to marking out from a single datum point. Work out the questions where drawing is required as accurately as possible and give short and clear answers to the questions. (When numbering lines remember that those which are identical should be given the same numbers.)

21 Figures A and B represent the end faces of a bright bar of 80 mm diameter. The centre has to be located.

a   Draw the lines in figure A where marking out is done with a centre square.

b   Draw the lines in figure B where odd-leg calipers are used.

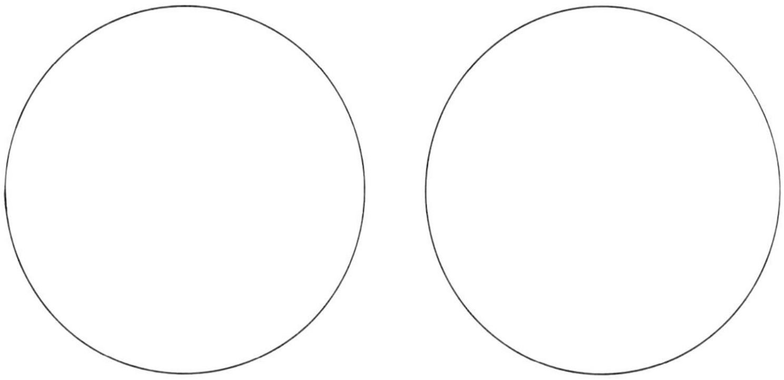

A   B

22 The figure shows a piece of thin sheet steel 102 mm square. Describe how to mark out:

a   the centre M of the plate

b   a circle of 100 mm diameter with centre located at M

c   an inscribed hexagon of that circle

Number the lines in marking out sequence, give a brief description of the procedure and list the equipment used.

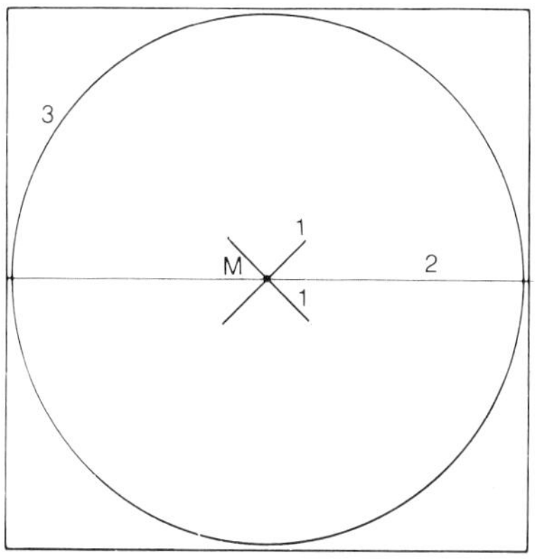

As an example, operations a and b have been completed below.

a   Draw the diagonals of the square using a straight edge and a scriber. Mark the centre point M with a centre punch.

b   Set the dividers to 50 mm and draw circle 3.

c   _____

_____

_____

23 The figure represents a piece of thin sheet steel 102 × 80 mm. Describe how to mark out:
a the centre M
b a circle with a diameter D of 100 mm
c an inscribed hexagon of that circle
Number the lines, give a brief description of the procedure and list the equipment used.

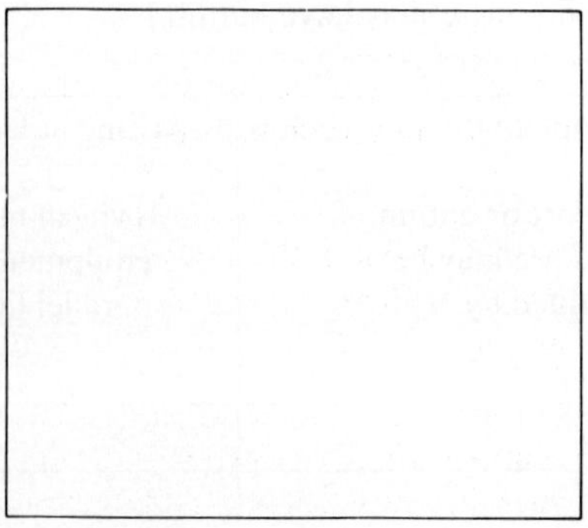

a _____

_____

_____

b _____

_____

c _____

_____

24 The circle in the figure represents a flange with diameter D of 450 mm. Describe how to:
a determine the centre
b scribe a pitch circle with a diameter D_p of 300 mm
c mark out 6 holes equally spaced on the pitch circle
Number the lines, give a brief description of the procedure and list the equipment used.

a _____

_____

_____

b _____

_____

c _____

_____

25 The circle in the figure represents a flange having a diameter D of 450 mm. Describe how to:
a determine the centre
b scribe a pitch circle with a diameter D_p of 300 mm
c mark out 4 holes equally spaced on the pitch circle
Number the lines, give a brief description of the procedure and list the equipment used.

_____

_____

Questions *26–35* relate to the figure shown here and refer to marking out from a single datum point and a single datum line. The drawing in this figure is of a plate from which is cut a workpiece which has been marked out already. The lines and points have been numbered, but not in the sequence in which they have been marked out. The correct answers to questions *26–34* should be indicated by circling the appropriate letters and/or numbers. Question *35* is a drawing task.

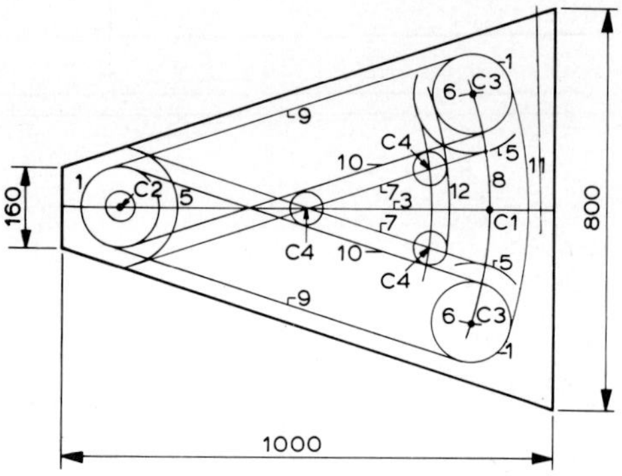

26 Indicate the number which identifies the line of
symmetry:
3 - 9 - 11 - 8

27 A centre mark has to be punched on the line of symmetry
to enable scribing arc 8; indicate which of the following is that mark:
C_1 - C_2 - C_3 - C_4

28 Indicate which one of the following is used to draw arc 8:
a   odd-leg calipers
b   dividers
c   a trammel

29 The centre mark C_1 is punched at the intersection of lines
3 and 8. Indicate which of the following is drawn from this mark:
a   the centre marks C_4
b   the arcs 6
c   the lines 11 and 12

30 Indicate from which of the following centre marks circles
1 are scribed with dividers:
C_1 - C_2 - C_3 - C_4

31 Indicate from which of the following centre marks the
arcs 5 are scribed with dividers:
C_1 - C_2 - C_3 - C_4

32 Indicate from which of the following centres marks the
arcs 11 and 12 are scribed with a trammel:
C_1 - C_2 - C_3 - C_4

33 Indicate which of the following must be drawn before
lines 9 can be drawn:
a   the lines 6
b   the lines 11 and 12
c   the arcs 5
d   the circles 1

34 Indicate to which of the following arcs or lines the lines 7 are tangential:
5 - 6 - 12 - 10

35 The figure shows the plate for questions *26–34* (increased
2½ times in size). As a start, line 1 and point C_2 are shown. Mark out all
lines, points and circles and number them in sequence. (Use the original
drawing, from which dimensions may be taken and increased 2½ times.)

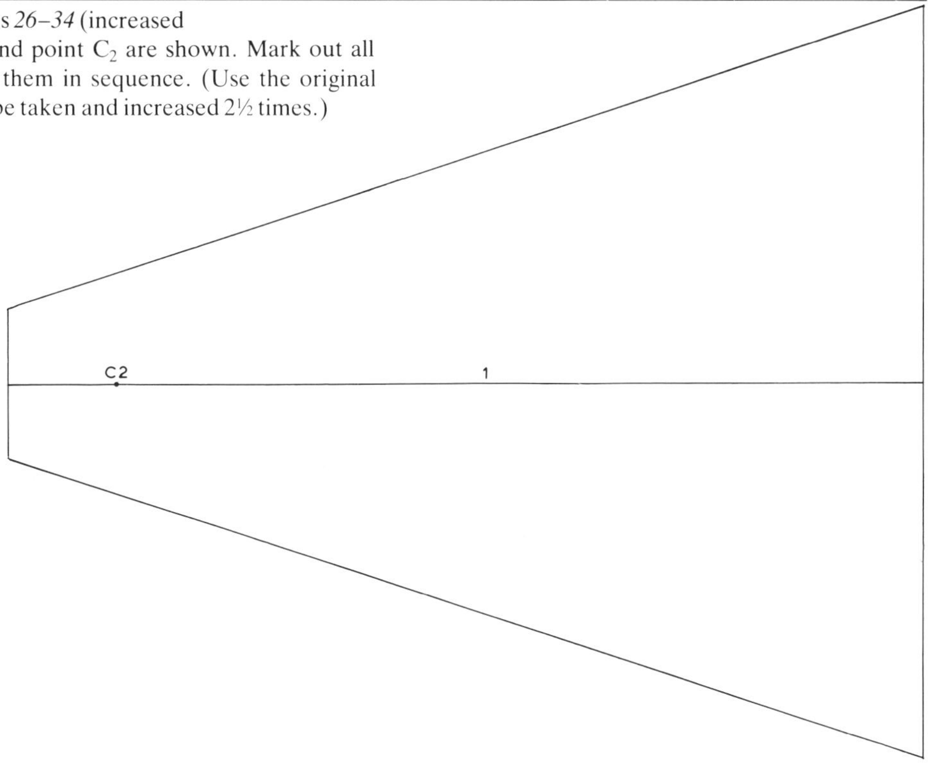

          © Stam Press Ltd. Amersham, 1986

Questions *36–44* refer to marking out from one or more datum faces.

36 Encircle the letter of the right answer. Datum faces are:
a identifiable because the dimensions are specified from those faces
b unworked faces
c those faces which are worked at the end to prevent damage
d always flat faces

37 Figure A is the drawing required for marking out workpiece B. The datum faces V₁ and V₂ are finished. Mark out B in accordance with the dimensions given in A. Write the marking out procedure in the space below.

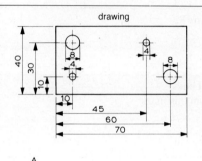

drawing

A

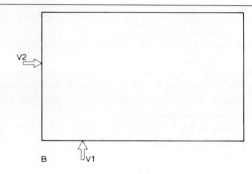

B

_____
_____
_____

38 Figure A is the working drawing for workpiece B which is machine finished on four edges. The hole has to be marked out for chain drilling the inside of the 28 mm square with a 5 mm drill. Mark it out and indicate the sequence of marking out by numbers. Write the marking out procedure in the space below.

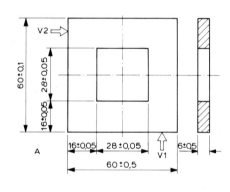

A

B

_____
_____
_____
_____
_____
_____
_____
_____
_____
_____

39 Questions *39–43* refer to the drawing on the right.
The workpiece is to be marked out on a surface plate.
  a   Indicate the three datum faces with arrows thus (→)
  b   Indicate these faces with V_1, V_2 and V_3, the largest
       area with V_1 and the smallest area with V_3.
Read questions *40–43* entirely through before commencing work on question *40*.

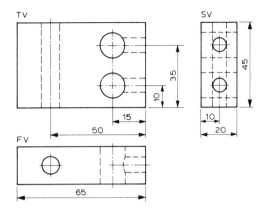

40 a   Draw the workpiece with the largest datum face rest-
           ing on the surface plate shown at right. A start has al-
           ready been made.
  b   Draw the lines which are marked out with a height
       gauge or a surface gauge scriber and mark them with
       1.
  c   Indicate on the drawing what the distance is between
       face V_1 and line 1.

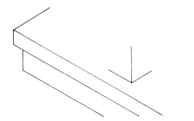

41 a   Draw the workpiece with its long narrow datum face
           resting on the surface plate.
  b   Draw all the lines which are to be marked out.
  c   Number the lower lines 2 and the upper lines 3.

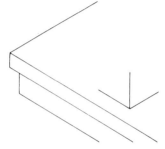

42 a   Draw the workpiece with its short narrow datum face
           resting on the surface plate.
  b   Draw all lines which are to be marked out.
  c   Number the lower lines 4 and the upper lines 5.
  d   Indicate on the drawing what the distance is between
       the datum face and line 4.

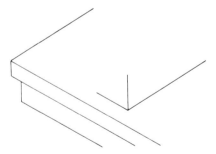

43 On the right the plan view of the workpiece in question *39* is shown, before
marking out.
  a   Draw the other TWO views.
  b   Draw line 1 in the front and side elevation.
  c   Draw the lines 2 and 3 on the plan and side elevation.
  d   Draw the line 4 on the plan.
  e   Draw the line 5 on the front elevation.
Number all lines.

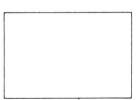

111   © Stam Press Ltd. Amersham, 1986

44 The sequence of marking out of a workpiece is shown on the right. Describe below how a series of twelve workpieces is to be marked out on the surface plate. As a distinction from dimensions, lines and points are denominated by numbers in italics.

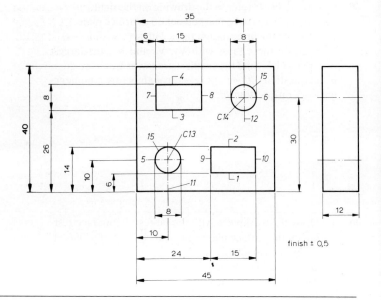

_____

_____

_____

_____

_____

_____

_____

_____

_____

45 A 960 mm square is to be marked out on material shaped as in the 1:20 scale drawing shown here. Using this scale, carry out the construction within the space of the drawing.

The procedure is as follows:
a Draw two lines 1 and 2 perpendicular to each other as near as possible to the bottom and left-hand edge using the try square. The intersection of the lines gives point 3.
b Mark out 960 mm from point 3 on the lines 1 and 2. Number these points 4 and 5.
c Draw a line 6 from 4 perpendicular to line 1 and a line 7 from 5 perpendicular to 2. The intersection of the lines 7 and 8 is point 8.
d Check the diagonals for equal length.

46 Again a 960 mm square is to be marked out on material shaped as in the 1:20 scale drawing shown here. Carry out the construction to the same scale, within the space of the drawing.

The procedure is as follows:
a Draw line 1 near to the bottom edge. Make a centre mark 2 on this line as near to the left edge as possible.
b Draw an arc of 750 mm from 2 on line 1 intersecting it at 3.
c Draw an arc of 1000 mm from 2 upward (arc 4) and an arc of 1250 mm from 3 to intersect arc 4. Number the intersection 5.
d Draw the line 2 to 5 and number this line 6.
e Mark out from 2 distances of 960 mm on lines 1 and 6. Number the intersections 7 and 8.
f From 7 and 8 mark out 960 mm again to find intersection 9. Draw the lines 7 to 9 and 8 to 9.
g Check the diagonals for equal length.

112

47

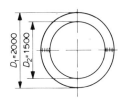

The ring shown has to be marked out on a piece of plate 2000 × 1000 mm. Draw in the space on the right half the ring to a scale of 1:25.
How many plates are required for one ring?

48

The ring described in question *47* may also be assembled as shown at the left. As an example a start has already been made in marking out the curved strips. Mark out all strips to a scale of 1:25.

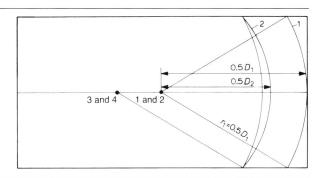

49 Compare questions *47* and *48* and answer the following questions:

  a Which method requires the most material?

  b Is the cutting length for the outer circle equal in each case?

  c Which method has the longest total cutting length?

  d Which method has the longest total welding length?

  e Which method requires the least material?

Questions *50–54* refer to the marking out, cutting and folding lines on sheet metal up to a thickness of 2 mm which can be folded so sharply that there is practically no internal bending radius. If the tolerance is 1 mm or more, the inside dimensions can be taken for the developed length. Colour the formed plate in each question.

50 Calculate the dimensions of the developed length and enter them on the drawing.

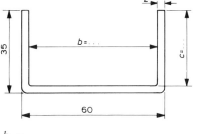

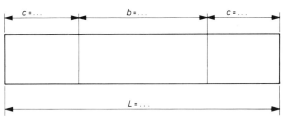

$b =$ _____

$c =$ _____

$L =$ _____

51 Calculate the dimensions of the developed length and enter them on the drawing.

$b =$

$c =$

$L =$

Why is one bending line drawn as a line of dashes?

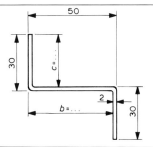

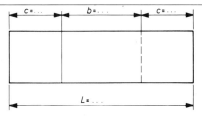

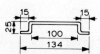

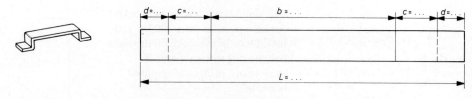

52 a Enter the dimensions in the development.

b Why are two bending lines shown as a line of dashes?

53 Figure A shows an end view of a component.

Figure B shows a plan view of its development. On the plan view enter the dimensions a b c and overall length L.

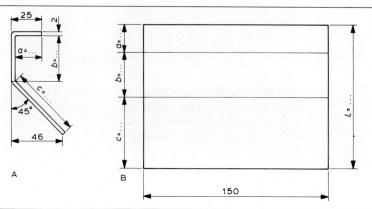

54

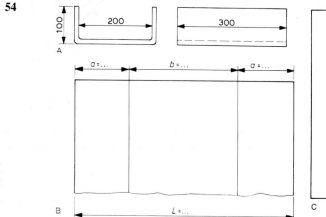

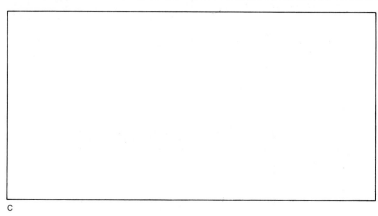

Figure A shows a 1 mm plate folded in two places. Find the dimensions of the development and enter the dimensions in B.

On C show how the largest possible number of plates B can be marked out on a plate measuring 2000 mm × 1000 mm. How many are there?

Questions 55–57 refer to marking out of cutting and folding lines on material with a thickness between 3 mm and 4 mm. The internal bending radius r_i is equal to or larger than the material thickness t. In these questions it is assumed that the neutral axis of the formed material lies in the middle of its thickness.
r_n = bend radius on the neutral axis.

55 In A colour the straight portions a_1 and b_1 and give portion c a different colour.

Complete the solution below:

$r_i =$ mm

$r_n = r_i + \frac{1}{2}t =$ +   =

$c = 1.57 \times r_n =$ ×   =

$L = a_1 = c = b_1 =$

           =

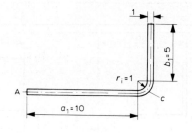

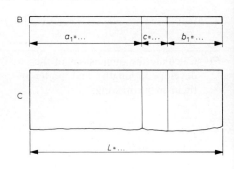

56   In A colour the straight portions a and b and give portion c a different colour.

Enter the dimension in the development (B = side elevation, C = plan).

Complete the solution below:

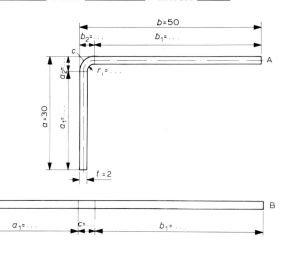

$a_2 = r_i = t =$ $b_1 = b - b_2$

$a_i = a - a^2 =$ $r_n = r_i + \frac{1}{2}t =$

$b_2 = r_i + t =$ $L = a_1 + c + b_1 =$

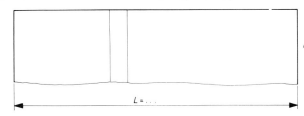

57   Give the portions a_{1L}, b_1 and a_{1R} the same colour. Give portions c a different colour. Find the dimensions for the development and enter them on the right. Then write them in the development.

$a_{2L} =$ $a_{2R} =$

$a_{1L} =$ $a_{1R} =$

$c_L =$ $c_R =$

$b_L =$ $b_R =$

$b_1 =$

$L =$

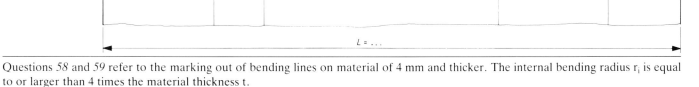

Questions 58 and 59 refer to the marking out of bending lines on material of 4 mm and thicker. The internal bending radius r_i is equal to or larger than 4 times the material thickness t.

The neutral axis of the bent material lies in the middle of its thickness.

r_n = the bend radius of the neutral axis.

58   Colour the cross-section A giving position c a different colour. Calculate the dimensions for the development (2 bending lines) and enter them below:

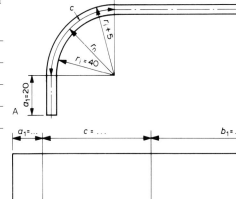

$r_i =$ _____

$r_n = r_i + \frac{1}{2}t =$ _____

$c = 1.57 \times r_n =$ _____

$L = a_1 + c + b_1 =$ _____

$L =$ _____

       © Stam Press Ltd. Amersham, 1986

59 Figure A is the cross-section of a gutter. Enter in B the dimensions of the development (2 bending lines for each bend).

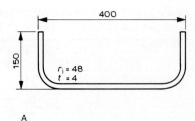

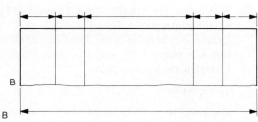

A

B

Questions *60–63* deal with location of points using coordinates.
P (3.2) means that P is determined by $x = 3$ and by $y = 2$.

---

60 Determine and mark:
P_1 (3.2) P_2 (8.2)

P_3 (3.7) P_4 (8.7)

Draw:
P_1P_2, P_3P_4, P_1P_3 and P_2P_4

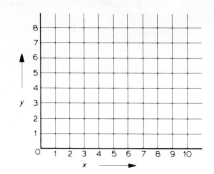

61 Determine and mark:
P_1 (5.1) P_2 (2.4)

P_3 (5.7) P_4 (8.4)

Draw:
P_2P_3, P_1P_4, P_3P_4 and P_1P_2

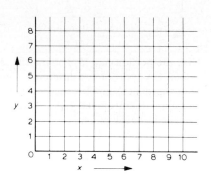

---

62 a Locate P_1 (2.1) and P_2 (10.1)

b Line P_1P_2 is the base of an equilateral triangle. The top of this triangle is P_3. Find the coordinate of P_3.
Instruction: Swing equal arcs from P_1 and P_2 with a radius P_1P_2. The intersection is the top P_3 of the equilateral tri-angle.

Bisect P_1P_2 to find its centre P_4.

$P_1P_2 = 8$ so $P_2P_4 = 4$

P_4 (........ , ...)

$P_1P_4 = 4$ and $P_1P_3 =$ ....

Therefore $P_3P_4 =$ ......$\sqrt{3}$ = ...

The x-value of P_3 = ...

The y-value of P_3 = ...

Therefore P_3 (........ , ...)

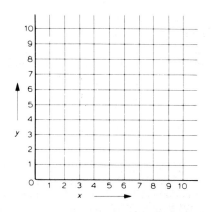

---

63 The centre of the circle shown is M (60.60). This means that the x-coordinate is 60 and the y-co-ordinate is 60. It is further given that the parts AB, BC and CA are equal. Draw AB and bisect it at P. Draw MP perpendicular to AB. Fill in:

MP = ... and   MA = $\frac{....}{2}$ = ...

PA = ...$\sqrt{}$........ = ... (why?)

Write the values found in the figure and fill in:

$x_A = 60 - 34.6$ = ...

$y_A = 60 + 20$ = ...

$x_B =$ ....   = ...

$y_B =$ ....   = ...

$x_C =$ ....   = ...

$y_C =$ ....   = ...

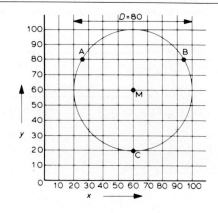

MEASURING Name: _____ Class: _____ Number: _____

The following questions span the syllabus subject matter and approximate to the level of those in the relevant examination paper of the City and Guilds of London Institute. Answers should be short and clear.

1 State the basic unit of length used in the UK. _____

2 State the internationally agreed temperature at which precise dimensions should be taken. _____

3 Name the principles upon which the operation of each of the following measuring instruments depends for its accuracy.

   a a vernier caliper _____

   b a dial gauge used as a comparator _____

   c a micrometer caliper _____

4 Name one instrument suitable for determining EACH of the following:

   a a linear dimension of 4.63 mm ± 0.01 mm _____

   b an angle of 20° ± 10′ _____

   c a linear dimension of 66.3 mm ± 0.1 mm _____

   d an angle of 45° ± 1° _____

   e the level of a flat surface _____

5 Explain why it is necessary to state a tolerance on precise dimensions. _____

6 Explain the difference between a unilateral and a bilateral tolerance. _____

7 A dimension is given as 56 ± 0.05 mm. State:

   a the basic dimension _____

   b the maximum dimension _____

   c the minimum dimension _____

   d the tolerance _____

   e whether the tolerance is unilateral or bilateral _____

8 Make a simple sketch to show how a plain circular bar may be checked for roundness.

9 Make a simple sketch to show how the inside and outside diameters of a plain bush may be checked for concentricity.

10 The figure below shows what is know as 'chain dimensioning'. State ONE disadvantage of this method of dimensioning.

_____

_____

_____

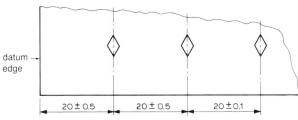

datum → edge

| 20 ± 0,5 | 20 ± 0,5 | 20 ± 0,1 |

dimensions are in millimetres

© Stam Press Ltd. Amersham, 1986

MEASURING   Name: _____ Class: _____ Number: _____

11 State the meaning of the following terms:
    a the RANGE of a dial test indicator
    b the ACCURACY OF DETERMINATION of a
       dimension
    c the BOTTOM LIMIT of a shaft diameter
    d the SENSITIVITY of a spirit level

12 a Explain the function of a protector slip gauge.
    b The table below shows the range, steps and the number of pieces inclu-
       ded in a metre set of slip gauges.
       Specify the slips selected to make up a dimension of 27.6325 mm using
       the minimum number of gauges.

Range mm	Steps mm	Pieces
1.0025	–	1
1.005	–	1
1.0075	–	1
1.01 to 1.49	0.01	49
0.05 to 9.50	0.50	19
10.00 to 50.00	10.00	5
75.00 and 100.00	–	2

13 State TWO reasons for marking out a component.

14 a Explain the meaning of the term 'a datum edge'.
    b Explain why tinned plate should not be marked using
       a scriber.
    c State THREE causes of inaccuracy when marking out
       with the aid of a steel rule.

15 Make neat sketches to show EACH of the **following**:
    a using a vernier height gauge to mark a line on a template
    b using a pair of odd leg calipers to draw a line 10 mm from the edge of a piece of plate 6m long.
    c using a surface gauge to set the face of a piece of material parallel to a surface table.

    a   b   c

16 Explain with the aid of a simple sketch THREE different ways of determining and marking the centre of a plain circular bar.

17 Name TWO different materials used for the manufacture
    of a surface plate.

18 Construct a circle of diameter 100 mm. On it mark the
centres of 6 holes equally spaced. Show your construction
lines faintly, but clearly.

19 Explain, with the aid of simple sketches, how the following may be marked out accurately on pieces of steel plate.
  a   Four holes of diameter 10 mm the centres of which are at the corners of a 60 mm square
  b   An arc of radius 500 mm subtending an arc of 30°
  c   A regular hexagon measuring 20 mm across flats

20 Explain with the aid of simple sketches how slip gauges and accessories may be used to enable accurate marking out to be performed.

BACKGROUND TO TECHNOLOGY

SUBJECT MATTER OF SECTIONS 1-8

SECTION 1
Basic Physical Quantities, Electricity and Magnetism

1. Introduction to the SI system
2. Structure and states of matter
3. Mass, force and weight
4. Mass per unit volume
5. Basic theory of electricity
6. Circuits
7. Magnetism

SECTION 2
Forces

1. Effects of force
2. Resultant and equilibrant forces
3. Resolution of forces
4. Moments; the theorem of moments
5. Conditions of equilibrium
6. Centre of gravity; equilibrium and stability
7. Friction

SECTION 3
Pressure

1. Pressure
2. Pressure exerted by liquids
3. Pressure in gases
4. Pressure on liquids
5. Connected vessels
6. Upthrust

SECTION 4
Heat

1. Heat and energy
2. Melting and solidifying
3. Evaporation and condensation
4. Dissolving and solidifying
5. Heat transfer

SECTION 5
Thermal Movement

1. Temperature
2. Thermal movement of solids
3. Thermal movement of liquids
4. Thermal movement of gases
5. The gas laws

SECTION 6
Motion

1. Linear motion at uniform velocity
2. Rotation at uniform speed
3. Direct transmission
4. Indirect transmission
5. Uniform acceleration from rest
6. Uniform acceleration and deceleration

SECTION 7
Energy

1. Force, mass and acceleration
2. Work
3. Power; rating and efficiency of machines
4. Potential and kinetic energy
5. Centripetal and centrifugal force

SECTION 8
Principles of tool construction; materials technology

1. Tools using the lever principle
2. Tools based on the pulley
3. Inclined plane and hydraulic equipment
4. Materials subject to tension and compression
5. Materials subject to shear

Note

All the books of 'Technology of Skilled Processes' are in one way or another related to the series 'Background to Technology'. In the case of 'Measuring and Marking Out' the sections 1.1, 5.1 and 5.2 must be studied and will be examined.

Information about the Syllabus and the books of Background to Technology can be obtained from The City and Guilds of London Institute, 76 Portland Place, Londen W1N4 AA or from the Publisher of these books Stam Press, Raans Road, Amersham, Bucks. HP6 6JJ.